Solutions Manual for

FOUNDATIONS OF COLLEGE CHEMISTRY
Eighth Edition
and
Fifth Alternate Edition

Solutions Manual for
FOUNDATIONS OF COLLEGE CHEMISTRY
Eighth Edition
and
Fifth Alternate Edition

Morris Hein

Susan Arena

Mount San Antonio College

Brooks/Cole Publishing Company
Pacific Grove, California

I(T)P ™ The trademark ITP is used under license.

Brooks/Cole Publishing Company
A Division of Wadsworth, Inc.

Printed in the United States of America

10   9   8   7   6   5   4   3   2

ISBN: 0-534-20007-9

TO THE STUDENT

This Solutions Manual contains answers to all questions and solutions to all problems at the end of each chapter in the text. Answers to the Review Exercises are not included.

This book will be valuable to you if you use it properly. It is not intended to be a substitute for answering the questions by yourself. It is important that you go through the process of writing down the answers to questions and the solutions to problems before you see them in the solutions manual. Once you have seen the answer, much of the value of whether or not you have learned the material is lost. This does not mean that you cannot learn from the answers. If you absolutely cannot answer a question or problem, study the answer carefully to see where you are having difficulty. You should spend sufficient time answering each assigned exercise. Only after you have made a serious attempt to answer a question or problem should you resort to the solutions manual.

Chemistry is one of the most challenging courses of study you will encounter. The core of the study of chemistry is problem solving. Skill at solving problems is achieved through effective and consistent practice. Watching and listening to others solve problems may be useful, but will not result in facility with chemistry problems. The methods used for solving problems in this manual are essentially the same as those used in the text. The basic steps in problem solving are fairly universal:

1. Read the problem carefully and determine the type of problem.
2. Develop a plan for solving the problem.
3. Write down the given information in an organized fashion.
4. Write a complete set-up and solution for the problem.
5. Use the solutions manual to check the answer and solution.

Your solution might be correct and yet will vary from the manual. Usually, these differences are the result of completing operations in a different order, or using separate steps instead of a single line approach.

If you make an error, take the time to analyze what went wrong. An error caused by an improper entry to the calculator can be frustrating, but it is not as serious as the inability to properly set up the problem. Do not become discouraged. Once you understand the steps in a problem, go back and rework it without looking at the answer.

The typing of this manual was done by Frank Arena. The legibility and clarity of the work is due in no small part to his efforts, as well as his patience and skill at the computer.

We have made every attempt to produce a manual with as few errors as possible. Please let us know of errors that you encounter, so that they can be corrected. Allow this manual to help you have success as you begin the great adventure of studying chemistry.

MORRIS HEIN
SUSAN ARENA
Mount San Antonio College

CONTENTS

CHAPTER 1

INTRODUCTION

1. No. The geocentric concept developed from incorrect interpretation of the observed apparent motion of the sun relative to the earth.

2. The principle goals of the alchemists were to find a method for prolonging human life and to change the base metals, such as iron and copper into gold.

3. The first use of the chemical balance in chemistry is analogous to the first use of the telescope in astronomy. These instruments made it possible to make measurements and observations with great precision, thus placing both sciences on a truly quantitative basis.

4. (a) Law   (e) Observation
   (b) Observation   (f) Theory
   (c) Hypothesis   (g) Hypothesis
   (d) Theory

5. The correct statements are a, b, d, f, g, h.

   (c) Knowledge of Western European people decreased markedly during the Dark Ages.
   (e) Most of the drugs and pharmaceuticals used in the United States today have been developed within the last century.
   (i) The two main branches of chemistry are organic and inorganic chemistry.

6. Answers will vary.

7. A theory is a well established hypothesis. It is an explanation based upon considerable evidence. A law, however, is a statement of a natural phenomenon without exception.

8. Reevaluate the hypothesis and perform further experiments.

9. Answers will vary.

10. $7 + 5 + 4 = 16$
    $6 - 2 + 8 = 12$
    $3 \times 8 \div 6 = 4$

11. Perkin discovered the dye mauve by chance. He investigated its properties and began the dye industry. Serendipity is enhanced by knowledge of the field. It is an important part of science because it may lead to whole new fields or discoveries.

12. Remsen and Fahlberg discovered saccharin, the first artificial sweetener, in 1878.

13. Tasting chemicals in the laboratory is not a safe practice because the toxicity of the chemicals may not be known, the purity of the chemical may be questionable, and the glassware used to make or store the chemical may be contaminated.

STANDARDS FOR MEASUREMENT

1. 10 dm = 1 m
   10,000 dm = 1 km

2. 25°C = 77°F   100°C = 212°F
   212°F − 77°F = 135°F

3. 7.6 cm

4. The volumetric flask is a more precise measuring instrument than the graduated cylinder. This is because the narrow opening at the point of measurement (calibration mark) in the neck of the flask means that a small error made in not filling exactly to the mark will make a smaller percentage error in total volume than will a similar error with the cylinder.

5. The three materials would sort out according to their densities with the most dense (mercury) at the bottom and the least dense (carbon tetrachloride) at the top. In the cylinder the solid magnesium would sink in the carbon tetrachloride and float on the liquid mercury.

6. Order of increasing density: ethyl alcohol, cottonseed oil, sulfur, salt, aluminum.

7. An argon-filled balloon would sink in a methane atmosphere because the density of argon is greater than that of methane.

8. The density of ice must be less than 0.926 g/mL and greater than 0.789 g/mL.

9. The principal advantages include:

   (a) conversion factors are multiples of 10, eliminating the necessity of fractions as measurements.

   (b) prefixes are the same throughout the metric system regardless of units. Therefore, there are fewer units to remember.

   (c) The metric system is used internationally.

   (d) The metric system uses mass, which remains constant, regardless of position (including international travel), while the American system uses weight, which changes relative to gravitational attraction.

10. (a)   gram = g   (g)   micrometer μm
    (b)   kilogram = kg   (h)   angstrom = Å
    (c)   milligram = mg   (i)   milliliter = mL
    (d)   microgram = μg   (j)   microliter = μL
    (e)   centimeter = cm   (k)   liter = L
    (f)   millimeter = mm

11. (a)   503   zero is significant
    (b)   0.007   zeroes are not significant
    (c)   4200   zeroes are not significant
    (d)   3.0030   zeroes are significant
    (e)   100.00   zeroes are significant
    (f) 8.00×10^2 zeroes are significant

12. Rule 1. When the first digit after those you want to retain is 4 or less, that digit and all others to its right are dropped. The last digit retained is not changed.

Rule 2. When the first digit after those you want to retain is 5 or greater, that digit and all others to the right of it are dropped and the last digit retained is increased by one.

13. Heat is a form of energy, while temperature is a measure of the intensity of heat (how hot the system is).

14. Density is the ratio of the mass of a substance to the volume occupied by that mass. Density has the units of mass over volume. Specific gravity is the ratio (no units) of the density of a substance to the density of a reference substance (usually water at a specific temperature for solids and liquids).

15. An individual is weighed in air, then seated on a chair suspended from a scale and lowered into a tank of warm water. After exhaling fully the individual submerges completely and is weighed while underwater.

16. Weight is not a measure of the amount of body fat, but rather the total composition of the body. Health risk is associated with an excess of fat in proportion to the total body. Leanness or obesity is indicated by the amount of fat compared to the rest of the body material.

17. Body density is determined by using the difference in weight in air and underwater. A lean person weighs more in water than an obese person and has a higher density as a result.

18. The correct statements are: a, c, d, e, g, h, i, j, l, n, p, q.

    (b)   The length 10 cm is equal to 100 mm.
    (f)   The sum of 32.276 + 2.134 should contain 5 significant figures.
    (k)   One centimeter is shorter than one inch.
    (m)   The number 0.0002983 in exponential notation is 2.983×10^{-4}
    (o)   Temperature is a measure of energy.

19. The number of significant figures (underlined) is given in parentheses:

(a) 0.0<u>25</u>   (2)
(b) <u>40.0</u>   (3)
(c) <u>22.4</u>   (3)
(d) 0.00<u>81</u>   (2)

(e) 0.0<u>404</u>   (3)
(f) <u>129.042</u>   (6)
(g) <u>5.50</u> $\times 10^3$ (3)
(h) <u>4.090</u> $\times 10^{-3}$ (4)

20. Rounding off to <u>three</u> significant figures:

(a) 93.2
(b) 8.87
(c) 0.0285
(d) 21.3

(e) 4.64
(f) 130. (1.30×10^2)
(g) 34.3
(h) 2.00 x 10^6

21. Exponential notation:

(a) 2.9×10^6
(b) 4.56×10^{-2}
(c) 5.8×10^{-1}
(d) 4.0822×10^3

(e) 8.40×10^{-3}
(f) 4.030×10^1
(g) 1.2×10^7
(h) 5.5×10^{-6}

22. (a) $\begin{array}{r} 12.62 \\ 1.5 \\ \underline{0.25} \\ 14.37 \end{array} = 14.4$

(b) $(4.68)(12.5) = 58.5$

(c) $(2.25 \times 10^3)(4.80 \times 10^4) = 10.8 \times 10^7 = 1.08 \times 10^8$

(d) $\dfrac{182.6}{4.6} = 39.7 = 4.0 \times 10^1$

(e) $\dfrac{(452)(6.2)}{14.3} = 196 = 2.0 \times 10^2$

(f) $\begin{array}{r} 1986. \\ 23.48 \\ \underline{0.012} \\ 2009.492 \end{array} = 2009$

(g) $(0.0394)(12.8) = 0.504$

(h) $(2.92 \times 10^{-3})(6.14 \times 10^5) = 17.9 \times 10^2 = 1.79 \times 10^3$

(i) $\dfrac{0.4278}{59.6} = 0.00718 = 7.18 \times 10^{-3}$

(j) $\dfrac{29.3}{(284)(415)} = 2.49 \times 10^{-4}$

23. Fractions to decimals (three significant figures)

(a) $\frac{5}{6} = 0.833$

(b) $\frac{3}{7} = 0.429$

(c) $\frac{12}{16} = 0.750$

(d) $\frac{9}{18} = 0.500$

24. (a) $3.42x = 6.5$

$$\frac{3.42x}{3.42} = \frac{6.5}{3.42}$$

$$x = \frac{6.5}{3.42} = 1.9$$

(b) $\frac{x}{12.32} = 7.05$

$$x = (7.05)(12.3) = 86.7$$

(c) $\frac{0.525}{x} = 0.25$

$$0.0525 = 0.25\ x$$

$$x = \frac{0.525}{0.25} = 2.1$$

(d) $0.298x = 15.3$

$$x = \frac{15.3}{0.298}$$

$$x = 51.3$$

(e) $\frac{x}{0.819} = 10.9$

$$x = (10.9)(0.819)$$

$$= 8.93$$

(f) $\frac{8.4}{x} = 282$

$$8.4 = 282\ x$$

$$x = \frac{8.4}{282} = 0.030$$

25. (a) $°C = \frac{212 - 32}{1.8}$

$°C = 100$

(b) $°F = 1.8(22) + 32$

$= 72°F$

(c) $K = 25 + 273$

$K = 298$

(d) $\frac{8.9\ g}{1\ mL} = \frac{40.90\ g}{V}$

$$(8.9\ g/mL)(V) = 40.90\ g$$

$$V = \frac{40.90\ g}{8.9\ g/mL} = 4.6\ mL$$

26. (a) $(28.0\ cm)\left(\frac{1\ m}{100\ cm}\right) = 0.280\ m$

(b) $(1000.\ m)\left(\frac{1\ km}{1000\ m}\right) = 1.000\ km$

(c) $(9.28\ cm)\left(\frac{10\ mm}{1\ cm}\right) = 92.8\ mm$

(d) $(150\ mm)\left(\frac{1\ m}{1000\ mm}\right)\left(\frac{1\ km}{1000\ m}\right) = 1.5 \times 10^{-4}\ km$

(e) $(0.606\ cm)\left(\frac{1\ m}{100\ cm}\right)\left(\frac{1\ km}{1000\ m}\right) = 6.06 \times 10^{-6}\ km$

(f) $(4.5\ cm)\left(\frac{1\ m}{100\ cm}\right)\left(\frac{1\text{Å}}{10^{-10}\ m}\right) = 4.5 \times 10^{8}\ \text{Å}$

(g) $(6.5 \times 10^{-7}\ m)\left(\frac{1\ \text{Å}}{10^{-10}\ m}\right) = 6.5 \times 10^{3}\ \text{Å}$

(h) $(12.1 \text{ m})\left(\frac{100 \text{ cm}}{1 \text{ m}}\right) = 1.21 \times 10^3 \text{ cm}$

(i) $(8.0 \text{ km})\left(\frac{1000 \text{ m}}{1 \text{ km}}\right) = 8.0 \times 10^3 \text{ m}$

(j) $(315 \text{ mm})\left(\frac{1 \text{ cm}}{10 \text{ mm}}\right) = 31.5 \text{ cm}$

(k) $(25 \text{ km})\left(\frac{1000 \text{ m}}{1 \text{ km}}\right)\left(\frac{1000 \text{ mm}}{1 \text{ m}}\right) = 2.5 \times 10^7 \text{ mm}$

(l) $(12 \text{ nm})\left(\frac{10^{-9} \text{ m}}{1 \text{ nm}}\right)\left(\frac{100 \text{ cm}}{1 \text{ m}}\right) = 1.2 \times 10^{-6} \text{ cm}$

(m) $(0.520 \text{ km})\left(\frac{1000 \text{ m}}{1 \text{ km}}\right)\left(\frac{100 \text{ cm}}{1 \text{ m}}\right) = 5.20 \times 10^4 \text{ cm}$

(n) $(3.884 \text{ Å})\left(\frac{10^{-10} \text{ m}}{1 \text{ Å}}\right)\left(\frac{1 \text{ nm}}{10^{-9} \text{ m}}\right) = 0.3884 \text{ nm}$

(o) $(42.2 \text{ in})\left(\frac{2.54 \text{ cm}}{1 \text{ in}}\right) = 107 \text{ cm}$

(p) $(0.64 \text{ mi})\left(\frac{5280 \text{ ft}}{1 \text{ mi}}\right)\left(\frac{12 \text{ in.}}{1 \text{ ft}}\right) = 4.1 \times 10^4 \text{ in.}$

(q) $(504 \text{ mi})\left(\frac{1.61 \text{ km}}{1 \text{ mi}}\right) = 811 \text{ km}$

(r) $(2.00 \text{ in}^2)\left(\frac{2.54 \text{ cm}}{1 \text{ in}}\right)^2 = 12.9 \text{ cm}^2$

(s) $(35.6 \text{ m})\left(\frac{100 \text{ cm}}{1 \text{ m}}\right)\left(\frac{1 \text{ in.}}{2.54 \text{ cm}}\right)\left(\frac{1 \text{ ft}}{12 \text{ in.}}\right) = 117 \text{ ft}$

(t) $(16.5 \text{ km})\left(\frac{1 \text{ mi}}{1.61 \text{ km}}\right) = 10.2 \text{ mi}$

(u) $(4.5 \text{ in.}^3)\left(\frac{2.5 \text{ cm}}{1 \text{ in.}}\right)^3\left(\frac{10 \text{ mm}}{1 \text{ cm}}\right)^3 = 7.4 \times 10^4 \text{ mm}^3$

(v) $(3.00 \text{ mi}^3)\left(\frac{1.61 \text{ km}}{1 \text{ mi}}\right)^3\left(\frac{10^6 \text{mm}}{1 \text{ km}}\right)^3 = 1.25 \times 10^{19} \text{ mm}^3$

27. (a) $(10.68 \text{ g})\left(\frac{1000 \text{ mg}}{1 \text{ g}}\right) = 1.068 \times 10^4 \text{ mg}$

(b) $(6.8 \times 10^4 \text{ mg})\left(\frac{1 \text{ g}}{1000 \text{ mg}}\right)\left(\frac{1 \text{ kg}}{1000 \text{ g}}\right) = 6.8 \times 10^{-2} \text{ kg}$

(c) $(8.54 \text{ g})\left(\frac{1 \text{ kg}}{1000 \text{ g}}\right) = 0.00854 \text{ kg}$

(d) $(42.8 \text{ kg})\left(\frac{2.2 \text{ lb}}{\text{kg}}\right) = 94.2 \text{ lb}$

(e) $(164 \text{ mg})\left(\frac{1 \text{ g}}{1000 \text{ mg}}\right) = 0.164 \text{ g}$

(f) $(0.65 \text{ kg})\left(\frac{1000 \text{ g}}{1 \text{ kg}}\right)\left(\frac{1000 \text{ mg}}{1 \text{ g}}\right) = 6.5 \times 10^5 \text{ mg}$

(g) $(5.5 \text{ kg})\left(\frac{1000 \text{ g}}{\text{kg}}\right) = 5.5 \times 10^3 \text{ g}$

(h) $(95 \text{ lb})\left(\frac{454 \text{ g}}{1 \text{ lb}}\right) = 4.3 \times 10^4 \text{ g}$

28. (a) $(25.0 \text{ mL})\left(\frac{1 \text{ L}}{1000 \text{ mL}}\right) = 2.50 \times 10^{-2} \text{ L}$

(b) $(22.4 \text{ L})\left(\frac{1000 \text{ mL}}{1 \text{ L}}\right) = 2.24 \times 10^4 \text{ mL}$

(c) $(3.5 \text{ qt})\left(\frac{946 \text{ mL}}{1 \text{ qt}}\right) = 3.3 \times 10^3 \text{ mL}$

(d) $(4.5 \times 10^4 \text{ ft}^3)\left(\frac{12 \text{ in}}{1 \text{ ft}}\right)^3\left(\frac{2.54 \text{ cm}}{1 \text{ in}}\right)^3\left(\frac{1 \text{ m}}{100 \text{ cm}}\right)^3 = 1.3 \times 10^3 \text{ m}^3$

(e) $(0.468 \text{ L})\left(\frac{1000 \text{ mL}}{1 \text{ L}}\right) = 468 \text{ mL}$

(f) $(35.6 \text{ L})\left(\frac{1 \text{ qt}}{0.946 \text{ L}}\right)\left(\frac{1 \text{ gal}}{4 \text{ qt}}\right) = 9.41 \text{ gal}$

(g) $(9.0 \text{ }\mu\text{L})\left(\frac{1 \text{ L}}{10^6 \mu\text{L}}\right)\left(\frac{1000 \text{ mL}}{\text{L}}\right) = 9.0 \times 10^{-3} \text{ mL}$

(h) $(20.0 \text{ gal})\left(\frac{4 \text{ qt}}{1 \text{ gal}}\right)\left(\frac{0.946 \text{ L}}{1 \text{ qt}}\right) = 75.7 \text{ L}$

29. (a) $(55 \text{ mi/hr})(1.61 \text{ km/mi}) = 89 \text{ km/hr}$

(b) $(55 \text{ mi/hr})(5280 \text{ ft/mi}) (1 \text{ hr}/3600 \text{ s}) = 81 \text{ ft/s}$

30. (a) $\left(\frac{100. \text{ m}}{8.9 \text{ s}}\right)\left(\frac{100 \text{ cm}}{\text{m}}\right)\left(\frac{1 \text{ in.}}{2.54 \text{ cm}}\right)\left(\frac{1 \text{ ft}}{12 \text{ in.}}\right) = 37 \text{ ft/s}$

(b) $\left(\frac{37 \text{ ft}}{1 \text{ s}}\right)\left(\frac{1 \text{ mi}}{5280 \text{ ft}}\right)\left(\frac{3600 \text{ s}}{1 \text{ hr}}\right) = 25 \text{ mi/hr}$

31. (24 students)(6.55 g/student) = 157 g NaCl
1.00 lb = 454 g
454 g - 157 g = 297 g NaCl remaining

32. (a) (27,000 mi/hr)(1 hr/3600 s) = 7.5 mi/s

    (b) (7.5 mi/s)(1.61 km/mi) = 12 km/s

33. (170 lb)(454 g/lb)(1 kg/1000 g) = 77.2 kg

34. (5.0 grains)(1 lb/7000 grains)(454 g/lb) = 0.32 g

35. $(9.3 \times 10^7$ mi)(1.61 km/mi)(10^5 cm/km)(1 s/3.00×10^{10} cm) = 5.0×10^2 s

36. (1 oz)(1 lb/16 oz)(454 g/1 lb)(1000 mg/g) = 3×10^4 mg

37. ($1.78/10 lb)(1 lb/0.454 kg) = $0.39/kg

38. ($345/oz)(14.58 oz/lb)(1 lb/454 g)(227 g) = $2520

39. (21 lb)(454 g/lb) = 9534 g condor
9534 g/3.2 g = 3.0×10^3 times heavier

40. ($0.35/L)(0.946 L/qt)(4 qt/gal)(15.8 gal) = $21

41. $\left(\dfrac{500. \text{ mi}}{34 \text{ mi/gal}}\right)$(4 qt/gal)(0.946 L/qt) = 56 L

42. (75 cm)(55 cm)(55 cm)(1 L/1000 cm^3) = 230 L

43. (20. drops/mL)(946 mL/qt)(4 qt/gal)(1 gal) = 7.6×10^4 drops

44. (42 gal)(4 qt/gal)(0.946 L/qt) = 160 L

45. (5.0 μL)(1L/10^6 μL)(1 mL/10^{-3} L) = 5.0×10^{-3} mL

46. (1.00 ft^3)(12 in./ft)^3(2.54 cm/in.)^3(1 mL/1 cm^3) = 2.83×10^4 mL

47. V = A x h   A = area   h = height

$A = V/h = \left(\dfrac{200 \text{ cm}^3}{0.5 \text{ nm}}\right)\left(\dfrac{1 \text{ nm}}{10^{-9} \text{ m}}\right)\left(\dfrac{1 \text{ m}}{100 \text{ cm}}\right)\left(\dfrac{1 \text{ m}}{100 \text{ cm}}\right)^2 = 4 \times 10^5$ m^2

48. $\dfrac{98.6 - 32}{1.8}$ = 37.0°C

49. (−100)(1.8) + 32 = −148°F
      −100°C is colder than −138°F

50. (a) $\dfrac{162 - 32}{1.8} = 72°C$ 　　　　　　　 (e) $1.8(32) + 32 = 90.°F$

(b) $\dfrac{0.0 - 32}{1.8} = -17.8°C$ 　　　　　　 (f) $\dfrac{-8.6 - 32}{1.8} = -22.6 °C$

(c) $\dfrac{0.0 - 32}{1.8} + 273 = 255.2 \ K$ 　　　 (g) $273 + 273 = 546 \ K$

(d) $1.8(-18) + 32 = -0.40°F$ 　　　　　 (h) $212 - 273 = -61 °C$

51. (a) 　$°F = °C$
　　　$°F = 1.8 \ (°C) + 32$
　　　$°F = 1.8(°F) + 32$
　　　$-32 = 0.8(°F)$
　　　$-32/0.8 = °F$
　　　$-40 = °F$

(b) 　$°F = -°C$
　　　$°F = 1.8(°C) + 32$
　　　$-°C = 1.8(°C) + 32$
　　　$2.8(°C) = -32$
　　　$°C = -32/2.8$
　　　$°C = -11.4$

52. $d = m/V = 78.26 \ g/50.00 \ mL = 1.565 \ g/mL$

53. $d = m/V = 39.9 \ g/12.8 \ mL = 3.12 \ g/mL$

54. $29.6 \ mL - 25.0 \ mL = 4.6 \ mL$ (V of chromium)
$d = m/V = 32.7 \ g/4.6 \ mL = 7.1 \ g/mL$

55. $d = m/V$
$m = dV = (1.19 \ g/mL)(500. \ mL) = 595 \ g$

56. $106.773 \ g - 42.817 \ g = 63.956 \ g$ (mass of the liquid)
$d = m/V = 63.956 \ g/50.0 \ mL = 1.28 \ g/mL$

57. $d = m/V$
$m = dV = (13.6 \ g/mL)(25.0 \ mL) = 3.40 \times 10^2 \ g$

58. $m = dV = (0.789 \ g/mL)(35.0 \ mL) = 27.6 \ g$ ethyl alcohol
$27.6 \ g + 49.28 \ g = 76.9 \ g$ (mass of cylinder and alcohol)

59. $d = m/V$ The cube with the largest V has the smallest D.
Use Table 2.5.

Cube A – lowest d　　　　　 1.74 g/mL – Mg
Cube B　　　　　　　　　　 2.70 g/mL – Al
Cube C – highest d　　　　　 10.5 g/mL – Ag

60. The volume of the Al **cube** is:

$V = m/d = (500. \ g)(2.70 \ g/mL) = 185 \ mL$
This is the same volume as the Au **cube** thus:

$m = dV = (185 \ mL)(19.3 \ g/mL) = 3.57 \times 10^3 \ g$

61. $d = m/V = 24.12 \ g/25.0 \ mL = 0.965 \ g/mL$

62. 150.50 g - 88.25 g = 62.25 g (mass of liquid)
d = m/V thus V = m/d = 62.25 g/1.25 g/mL = 49.8 mL
The container must hold 50 mL.

63. H_2O 50 g/(1.0 g/mL) = 50 mL
alcohol   50 g/(0.789 g/mL) = 60 mL
alcohol has the greater volume

64. V = (2.00 cm)(15.0 cm)(6.00 cm)(1 mL/1 cm^3) = 180. mL
d = m/V = 3300 g/180 mL = 18.3 g/mL
The density of gold is 19.3 g/mL from Table 2.5, therefore, the gold bar is not pure gold, since its density is only 18.3 g/mL, or it is hollow inside.

65. $\left(\dfrac{9.33 \text{ x } 10^4 \text{ g}}{19.3 \text{ g/ml}} \right)$ = 4.83 x 10^3 mL = 4.83 x 10^3 cm^3

(9.33 x 10^4 g)(1 lb/454 g)(14.58 oz/1 lb) = (3.00 x 10^3 oz)

(3.00 x 10^3 oz)($345/oz) = $1,033,718

66. Volume of slug   30.7 mL - 25.0 mL = 5.7 mL

Density of slug d = m/V = 15.434 g/5.7 mL = 2.7 g/mL
Mass of liquid, cylinder, and slug   125.934 g
Mass of slug (subtract)   −15.434 g
Mass of cylinder (subtract)   −89.450 g
Mass of the liquid   21.050 g

Density of liquid d = m/V = 21.050 g/25.0 mL = 0.842 g/mL

CLASSIFICATION OF MATTER

1. Answers will vary

2. (a)   The attractive forces among the ultimate particles of a solid (atoms, ions, or molecules) are strong enough to hold these particles in fixed position within the solid and thus maintain the solid in a definite shape. The attractive forces among the ultimate particles of a liquid (usually molecules) are sufficiently strong enough to hold them together (preventing the liquid from rapidly becoming a gas) but are not strong enough to hold the particles in fixed positions (as in a solid).

    (b)   The ultimate particles in a liquid are quite closely packed (essentially in contact with each other) and thus the volume of the liquid is fixed at a given temperature. But, the ultimate particles in a gas are relatively far apart and essentially independent of each other. Consequently, the gas does not have a definite volume.

    (c)   In a gas the particles are relatively far apart and are easily compressed, but in a solid the particles are closely packed together and are virtually incompressible.

3. The water in the beaker does not fill the test tube. Since the tube is filled with air (a gas) and two objects cannot occupy the same space at the same time, the gas is shown to occupy space.

4. Mercury and water are the only liquids in the table which are not mixtures.

5. Air is the only gas mixture found in the table.

6. Three phases are present within the bottle; solid and liquid are observed visually, while gas is detected by the immediate odor.

7. The system is heterogeneous as multiple phases are present.

8. A system containing only one substance is not necessarily homogeneous. Two phases may be present. Example: ice in water.

9. A system containing two or more substances is not necessarily heterogeneous. In a solution only one phase is present. Examples: sugar dissolved in water, dilute sulfuric acid.

10. Silicon   25.67%   Hydrogen   0.87%

    In 100 g $\dfrac{25.67 \text{ g Si}}{0.87 \text{ g H}} = 30 \text{ Si}/1 \text{ g H}$

    Si is 28 times heavier than H, thus since $30 > 28$, there are more Si atoms than H atoms.

11. The symbol of an element represents the element itself. It may stand for a single atom or a given quantity of the element.

12.
phosphorus	P	sodium	Na
aluminum	Al	nitrogen	N
hydrogen	H	nickel	Ni
potassium	K	silver	Ag
magnesium	Mg	plutonium	Pu

13. (a) Si – 1 atom silicon   SI – System International or 1 atom sulfur
                                      1 atom iodine

    (b) Pb – 1 atom lead   PB – 1 atom phosphorus   1 atom boron

    (c) 4P – 4 atoms phosphorus P_4 – 1 molecule phosphorus
                                     (made of 4 phosphorus atoms)

14.
Na	sodium	Ag	silver
K	potassium	W	tungsten
Fe	iron	Au	gold
Sb	antimony	Hg	mercury
Sn	tin	Pb	lead

15.
H	hydrogen	S	sulfur
B	boron	K	potassium
C	carbon	V	vanadium
N	nitrogen	Y	yttrium
O	oxygen	I	iodine
F	fluorine	W	tungsten
P	phosphorus	U	uranium

16. In an element all atoms are alike, while a compound contains two or more elements (different atoms) which are chemically combined.   Compounds may be decomposed into simpler substances while elements cannot.

17. 84 metals   7 metalloids   18 nonmetals

18. 7 metals   1 metalloid   2 nonmetals

19. 1 metal   0 metalloids   5 nonmetals

20. The symbol for gold is based upon the Latin word for gold, aurum.

21. (a) iodine
    (b) bromine

22. A compound is composed of two or more elements which are chemically combined in a definite proportion by mass. Its properties differ from those of its components.   A mixture is the physical combining of two or more substances (not necessarily elements). The composition may vary, the substances retain their properties, and it may be separated by physical means.

23. Molecular compounds exist as molecules formed from two or more elements bonded together. Ionic compounds exist as cations and anions held together by electrical attractions.

24. Compounds are distinguished from one another by their characteristic properties.

25. (a) potassium, iodine
(b) sodium, carbon, oxygen
(c) aluminum, oxygen
(d) calcium, bromine
(e) carbon, chlorine
(f) magnesium, bromine
(g) hydrogen, nitrogen, oxygen
(h) barium, sulfur, oxygen
(i) aluminum, phosphorus, oxygen
(j) hydrogen, carbon, oxygen

26. (a) ZnO   (e) CaF_2
(b) $KClO_3$ (f) $PbCrO_4$
(c) NaOH   (g) C_2H_6O
(d) $AlBr_3$ (h) C_6H_6

27. (a) 2 atoms H, 1 atom O
(b) 1 atom Al, 3 atoms Br
(c) 2 atoms Na, 1 atom S, 4 atoms O
(d) 1 atom Ni, 2 atoms N, 6 atoms O
(e) 12 atoms C, 22 atoms H, 11 atoms O

28. (a) 2 atoms   (f) 8 atoms
(b) 5 atoms   (g) 5 atoms
(c) 2 atoms   (h) 17 atoms
(d) 9 atoms   (i) 16 atoms
(e) 11 atoms

29. (a) H_2 – 2 atoms
(b) H_2O – 3 atoms
(c) H_2SO_4 – 7 atoms

30. Cations are positively charged, while anions are charged negatively.

31. H_2 – hydrogen Cl_2 – chlorine
N_2 – nitrogen Br_2 – bromine
O_2 – oxygen I_2 – iodine
F_2 – fluorine

32. (a) 1 atom O   (b) 4 atoms O   (c) 2 atoms O
(d) 3 atoms O   (e) 9 atoms O

33. (a) 2 atoms H   (c) 12 atoms H
(b) 6 atoms H   (d) 4 atoms H

34. Homogeneous mixtures contain only one phase, while heterogeneous mixtures contain two or more phases.

35. (a) mixture   (e) mixture
    (b) element   (f) element
    (c) compound   (g) compound
    (d) element   (h) mixture

36. (a) mixture   (e) compound
    (b) compound   (f) element
    (c) element   (g) mixture
    (d) mixture   (h) compound

37. The substance is a compound. It decomposed into simpler substances with properties different than the original substance.

38.

Metals	Nonmetals
solid at room T (except Hg)	solids, liquids or gas at room T
luster	dull (no luster)
conduct heat & electricity	insulator (do not conduct electricity
malleable	react with each other forming compounds
react with nonmetals to form compounds	

39. diatomic molecules   a, c, e

40. Two allotropes of carbon are graphite and diamond. Diamond has distinct octahedral crystals which are colorless when pure. Diamond is the hardest known substance and does not conduct heat or electricity. Graphite is composed of layers of carbon atoms which slip easily over one another. Graphite is an excellent conductor of electricity.

41. Charcoal – destructive distillation of wood.
Bone black – destructive distillation of bones or wastes.
Carbon black – residue from burning natural gas.

42. Uses for graphite include: lubricants, "lead" in pencils, electrodes in dry cells, paints, water purification systems, gas masks, decolorizing, carbon paper, printer's ink, and tires.

43. Answers will vary. Several examples are given.

Petroleum – fuel, source of hydrocarbons for polymers, paints, and solvents.
Carbon dioxide – greenhouse gas, released in cellular respiration and used by plants in photosynthesis.
Sugars – nutrient for living organisms

44. Correct statements are: c, f, g, j, m, o, q, t, u, w, x, y, aa, dd.

    (a)   Gases are the least compact state of matter.

    (b)   Liquids have a definite volume but no definite shape.

    (d)   Wood is heterogeneous.

(e)　Wood is a mixture.

(h)　Many systems made of only one substance are homogeneous.

(i)　Some systems containing two or more substances are heterogeneous.

(k)　The smallest unit of an element that can exist and enter into a chemical reaction is called an atom.

(l)　The basic building blocks of all substances, which cannot be decomposed into simpler substances by ordinary chemical change, are elements.

(n)　The most abundant element in the human body, by mass, is oxygen.

(p)　The symbol for copper is Cu.

(r)　The symbol for potassium is K.

(s)　The symbol for lead is Pb.

(v)　The smallest uncharged individual unit of a compound formed by the union of two or more atoms is called a molecule.

(z)　A general property of metals is that they are good conductors of heat and electricity.

(bb)　Malleable means a substance will bend or shape when it is struck a hard blow.

(cc)　Elements that have properties intermediate between metals and nonmetals are called metalloids.

45. mass of CCl_4　$m = dV = (1.595 \text{ g/mL})(2.50 \text{ mL}) = 3.99 \text{ g}$

mass of CBr_4　$m = dV = (3.420 \text{ g/mL})(3.50 \text{ mL}) = 12.0 \text{ g}$

total mass of mixture　$3.99 \text{ g} + 12.0 \text{ g} = 16.0 \text{ g}$

density of solution　$d = m/V = \dfrac{16.0 \text{ g}}{2.50 \text{ mL} + 3.50 \text{ mL}} = 2.67 \text{ g/mL}$

46. mass of Hg　$m = dV = (13.6 \text{ g/mL})(12.5 \text{ mL}) = 1.70 \times 10^2 \text{ g}$

volume of Br_2　$V = m/d = \dfrac{1.70 \times 10^2 \text{g}}{3.12 \text{ g/mL}} = 54.5 \text{ mL}$

47.　60. g Au　　　　　　　$(60. \text{ g}/80. \text{ g})(100) = 75\% \text{ Au}$
　　　8.0 g Ag
　　　12. g Cu　　　　　　　$(0.75)(24 \text{ carat}) = 18 \text{ carat}$
　　　80. g alloy

48.　12 g C　　　　　　　$\left(\dfrac{12 \text{ g}}{16 \text{ g}}\right)(100) = 75\% \text{ C}$
　　　4 g H
　　　16 g

49.　$(0.90)(8420 \text{ g}) = 7.6 \times 10^3 \text{ g}$　Au

50.　$450 \text{ kg} = 0.25(x)$　　　　　$1180 \text{ kg} = 0.75(x)$

1800 kg alloy $= x$　if all Ni used　　1570 kg alloy $= x$　if all Cu used

Only 1570 kg of alloy can be manufactured before running out of copper

CHAPTER 4
PROPERTIES OF MATTER

1. Solid

2. Solid   102 K = –171°C   mp is –101.6°C

3. Droplets of silver colored mercury appear near the mouth of the test tube.

4. Magnesium and oxygen form white solid magnesium oxide.

5. In the electrolysis of water, the liquid is converted to hydrogen and oxygen gases.

6. Physical properties are characteristics which may be determined without altering the composition of the substance.   Chemical properties describe the ability of a substance to form new substances by reaction or decomposition.

7. A new substance is always formed during a chemical change, but never formed during physical changes.

8. (a) physical   (d) chemical
   (b) physical   (e) physical
   (c) physical   (f) chemical

9. (a) chemical   (d) chemical
   (b) physical   (e) chemical
   (c) physical   (f) physical

10. Although the appearance of the platinum wire changed during the heating, the original appearance was restored when the wire cooled.   No change in the composition of the platinum could be detected.

11. The copper wire, like the platinum wire, changed to a glowing red color when heated (physical change).   Upon cooling, the original appearance of the copper wire was not restored, but a new substance, black copper(II) oxide, had appeared in its stead.

12. When mercury(II) oxide is heated, the red powder changes color and produces silvery liquid globules of mercury and oxygen gas.   Evidence of the oxygen is observed when the gas given off from the heating is observed to enhance the burning of a wood splint.

13. (a) Reactants: copper and oxygen; Product: copper(II) oxide

    (b) Reactant: mercury(II) oxide; Products: mercury and oxygen

14. In a chemical change, any change of mass that occurs is so small as to be undetectable in an experiment, though it can be calculated from the energy change. Since it is undetectable, we say that mass is neither lost nor gained.

15. Potential energy is the energy of position. By the position of an object, it has the potential of movement to a lower energy state. Kinetic energy is the energy matter possesses due to its motion.

16. The kinetic energy is converted to thermal energy (heat), chiefly in the brake system, and eventually dissipated into the atmosphere.

17. The tranformation of kinetic energy to thermal energy (heat) is responsible for the fiery reentry of a space vehicle.

18. The energy equivalent of mass, as given by Einstein's expression, energy = (mass)(speed of light)^2, shows that a huge amount of energy is released when any measurable amount of matter is converted to energy. The mass equivalent of the energy released by a bulb flashing is too small to be detected by ordinary laboratory methods.

19. Foods are assigned an energy value by burning them in a calorimeter and measuring the heat released. Since the products of combustion and metabolism are the same, the caloric content of food is assigned by this method.

20. The hot pack contains a solution of sodium acetate or sodium thiosulfate. A small crystal is added by squeezing a corner of the bag or by bending a small metal activator. The solution crystallizes and heat is released to the surroundings. To reuse it, the pack is heated in boiling water until the crystals dissolve. Then, it is slowly cooled and stored till needed.

21. The correct statements are a, d, f, h, i.

    (b) When heated in the air, a platinum wire has a constant mass.

    (c) When heated in the air, a copper wire gains mass.

    (e) Boiling water represents a physical changes because a change of state occurs without a change in composition.

    (g) All of the following represent physical changes: breaking a stick, melting wax, folding a napkin.

    (j) A stretched rubber band possesses potential energy.

22. (a) $118.0°C + 273 = 391.0$ K

    (b) $(118.0°C)1.8 + 32 = 244.4°F$

23.
15.0 g	Fe
16.0 g	S
18.5 g	sand
49.5 g	total

    $\left(\dfrac{15.0 \text{ g Fe}}{49.5 \text{ g}}\right)(100) = 30.3\%$ Fe

24. $(1.80 \text{ g Cu})\left(\dfrac{1.251 \text{ g copper(II) oxide}}{1.00 \text{ g Cu}}\right) = 2.25 \text{ g copper (II) oxide}$

25. From Section 4.3 we see 1.00 g Cu combines with 0.251 g oxygen

$\left(\dfrac{1.00 \text{ g Cu}}{0.251 \text{ g oxygen}}\right)(5.50 \text{ g oxygen}) = 21.9 \text{ g Cu}$

26. $(65.0 \text{ g mercury (II) oxide})\left(\dfrac{92.6 \text{ g Hg}}{100 \text{ g mercury(II) oxide}}\right) = 60.2 \text{ g Hg}$

27. $\left(\dfrac{3.40 \text{ g fat}}{40.0 \text{ g}}\right)(100) = 8.50\% \text{ fat}$

28. magnesium $+$ oxygen $\longrightarrow$ magnesium oxide
     10.5 g x 17.4 g

    (a)   Law of Conservation of Mass: mass of reactants = mass of products

$$10.5 + x = 17.4 \text{ g}$$
$$x = 17.4 \text{ g} - 10.5 \text{ g} = 6.9 \text{ g oxygen}$$

    (b) $\left(\dfrac{10.5 \text{ g}}{17.4 \text{ g}}\right)(100) = 60.3\% \text{ Mg}$

29. aluminum $+$ chlorine $\longrightarrow$ aluminum chloride
     1.00 g   4.94 g

mass of chlorine $= 4.94 \text{ g} - 1.00 \text{ g} = 3.94 \text{ g}$

$\left(\dfrac{3.94 \text{ g chlorine}}{1.00 \text{ g aluminum}}\right)(5.50 \text{ g aluminum}) = 21.7 \text{ g chlorine}$

30. $E = (m)(\text{specific heat})(\Delta t)$

    $= (80. \text{ g})(4.184 \text{ J/g}°\text{C})(70.°\text{C} - 20.°\text{C})$

    $= 1.7 \times 10^4 \text{ J}$

31. $E = (m)(\text{specific heat})(\Delta t)$

    $= (80. \text{ g})(0.473 \text{ J/g}°\text{C})(70.°\text{C} - 20.°\text{C})$

    $= 1.9 \times 10^3 \text{ J}$

32. $E = (m)(\text{specific heat})(\Delta t)$

specific heat $= \dfrac{E}{(m)(\Delta t)} = \dfrac{5.866 \times 10^3 \text{ J}}{(250. \text{ g})(100.°\text{C} - 22°\text{C})} = 0.30 \text{ J/g }°\text{C}$

33. energy gained by the water $=$ (m)(specific heat)(Δ t)

$$= (100. \text{ g})(4.184 \text{ J/g } °C)(29.0°C - 25.0°C)$$

$$= 1.7 \text{ x } 10^3 \text{ J}$$

energy lost by the metal $=$ energy gained by the water $= 1.7 \text{ x } 10^3 \text{ J}$

$E = $ (m)(specific heat)(Δ t)

specific heat $= \dfrac{E}{(m)(\Delta t)} = \dfrac{1.7 \text{ x } 10^3 \text{J}}{(20.0 \text{ g})(174°C)} = 0.49 \text{ J/g } °C$

34. heat lost $=$ heat gained $\quad\quad\quad x =$ final temperature

(m)(specific heat)(Δ t) $=$ (m)(specific heat)(Δ t)

(10. g)(1.000 cal/g $°C$)($50.°C - x$) $=$ (50. g)(1.000 cal/g $°C$)($x - 10.°C$)

$500 \text{ cal} - 10x \text{ cal/°C} = 50x \text{ cal/°C} - 500 \text{ cal}$

$1000 \text{ cal} = 60x \text{ cal/°C}$

$x = 17°C$

35. heat lost by gold $=$ heat gained by water $\quad\quad x =$ final temperature

(m)(specific heat)(Δ t) $=$ (m)(specific heat)(Δ t)

(325 g)(0.131 J/g $°C$)($427°C - x$) $=$ (200. g)(4.184 J/g $°C$)($x - 22.0°C$)

$18179 \text{ J} - 42.575x \text{ J/°C} = 836.8x \text{ J/°C} - 18409.6 \text{ J}$

$36588 \text{ J} = 879.375x \text{ J/°C}$

$41.6 °C = x$

36. $E = $ (m)(specific heat)(Δ t)

$$= (250. \text{ g})(0.096 \text{ cal/g } °C)(150.°C - 24°C)$$

$$= 3.0 \text{ x } 10^3 \text{ cal}$$

37. $E = $ (m)(specific heat)(Δ t)

$40,000. \text{ J} = (500. \text{ g})(4.184 \text{ J/g } °C)(x - 10.0°C)$

$40,000. \text{ J} = 2092x \text{ J/°C} - 20920 \text{ J}$

$60,920 \text{ J} = 2092x \text{ J/°C}$

$29.1°C = x$

38. $E = $ (m)(specific heat)(Δ t)

specific heat $= \dfrac{E}{m(\Delta t)} = \dfrac{3.07 \text{ x } 10^4 \text{J}}{(1.00 \text{ x } 10^3 \text{g})(630.°C - 20.°C)}$

$= 5.03 \text{ x } 10^{-2} \text{ J/g } °C$

39. Specific heats for the metals are Fe: 0.473 J/g°C; Cu: 0.385 J/g°C; Al: 0.900 J/g°C. The metal with the lowest specific heat will warm most quickly, therefore the copper pan heats fastest and fries the egg fastest.

40. heat lost by iron = heat gained by water x = final temperature

$$(500. \text{ g})(0.473 \text{ J/g °C})(212°C - x) = (2.0 \times 10^3 \text{ g})(4.184 \text{ J/g °C})(x - 24.0°C)$$

$$50138 \text{ J} - 236.5x \text{ J/°C} = 8368x \text{ J/°C} - 200832 \text{ J}$$

$$250970 \text{ J} = 8604.5x \text{ J/°C}$$

$$x = 29°C$$

$$\Delta t = 29°C - 24°C = 5°C$$

41.   heat lost by coal = heat gained by water x = mass of coal

$$(5500 \text{ cal/g})x = (500. \text{ g})(1.00 \text{ cal/g °C})(90.°C - 20.°C) = 35,000 \text{ cal}$$

$$x = 6.36 \text{ g} = 6.4 \text{ g coal}$$

42. (a) 1 μg mass = 9.0×10^7 J = 1 x 10^{-6} g mass

$$(5.5 \text{ g})\left(\frac{9.0 \times 10^7 \text{ J}}{1 \times 10^{-6} \text{ g}}\right) = 5.0 \times 10^{14} \text{ J}$$

$$(5.0 \times 10^{14}\text{J})\left(\frac{1 \text{ cal}}{4.184 \text{ J}}\right) = 1.2 \times 10^{14} \text{ cal}$$

(b) $(5.0 \times 10^{14} \text{ J})\left(\frac{1 \text{ gal}}{1.27 \ 10^6 \text{ J}}\right) = 3.9 \times 10^8 \text{ gal}$

EARLY ATOMIC THEORY AND STRUCTURE

1.   Element   Atomic number

   (a) copper   29
   (b) nitrogen   7
   (c) phosphorous   15
   (d) radium   88
   (e) zinc   30

2. The neutron is about 1840 times heavier than an electron.

3. Gold nuclei are very massive (compared to an alpha particle) and have a large positive charge. As the alpha particles approach the atom, some are deflected by this positive charge. Those approaching a gold nucleus directly are deflected backwards by the massive positive nucleus.

4. Particle   charge   mass

    proton   + 1   1 amu
    neutron   0   1 amu
    electron   −1   0

5. In the atom, protons and neutrons are found within the nucleus. Electrons occupy the remaining space within the atom outside the nucleus.

6. The nucleus of an atom contains nearly all of its mass.

7. (a)   The nucleus of the atom contains most of the mass since only a collision with a very dense, massive object would cause an alpha particle to be deflected back towards the source.

    (b)   The deflection of the alpha particles from their initial flight indicates the nucleus of the atom is positively charged.

    (c)   Most alpha particles pass through the gold foil undeflected leading to the conclusion that the atom is mostly empty space.

8. (a)   Dalton contributed the concept that each element is composed of atoms which are unique, and can combine in ratios of small whole numbers.

    (b)   Thomson discovered the electron, determined its properties, and found the mass of a proton is 1840 times the mass of the electron. He developed the Thomson model of the atom.

    (c)   Rutherford devised the model of a nuclear atom with a positive charge and mass concentrated in the nucleus. Most of the atom is empty space.

9. The correct statements are a, b, c, f, g.

   (d) Chadwick stated that atoms are composed of protons, neutrons and electrons.

   (e) Portions of Dalton's theory are still considered valid today.

   (h) The nucleus of an atom contains protons and neutrons

10.

	substance	protons	neutrons	electrons
(a)	$^{79}_{35}$Br	35	44	35
(b)	$^{131}_{56}$Ba	56	75	56
(c)	$^{238}_{92}$U	92	146	92
(d)	$^{56}_{26}$Fe	26	30	26

11. Isotopic notation $^{A}_{Z}$X

   Z represents the atomic number
   A represents the mass number

12. (a) $^{55}_{26}$Fe   (d) $^{29}_{14}$Si

    (b) $^{26}_{12}$Mg   (e) $^{188}_{79}$Au

    (c) $^{6}_{3}$Li

13. (a) $^{59}_{27}$Co   (b) $^{184}_{74}$W

14. The correct statements are b, c, f, g, h, k.

   (a) An element with an atomic number of 29 has 29 protons and 29 electrons.

   (d) An atom of $^{31}_{15}$P contains 15 protons, 16 neutrons, and 15 electrons.

   (e) In the isotope of $^{6}_{3}$Li Z = 3 and A = 6

   (i) $^{24}_{11}$Na has one more neutron than $^{23}_{11}$Na

   (j) $^{24}_{11}$Na has one more neutron than $^{23}_{11}$Na

   (l) Most elements exist in nature as mixtures of isotopes.

15. Atomic masses are not whole numbers because:

   (a) the neutron and proton do not have identical masses and neither is exactly 1 amu.

   (b) most elements exist in nature as a mixture of isotopes. The atomic mass is the average of all these isotopes.

16. The isotope of C with a mass of 12 is an exact number by definition. The mass of other isotopes such as $^{63}_{29}$Cu will not be an exact number for reasons given in Exercise 15.

17. The most abundant isotope is 40. This is certain as 40 is the average of all isotopes and is lowest on the list. An arithmetic average would be between 40 and 48. Since the atomic mass is 40.08, there must be only small amounts of the other isotopes.

18. Isotopes contain the same number of protons and the same number of electrons. Isotopes have different numbers of neutrons and thus different atomic masses.

19. The isotopes of hydrogen are protium, deuterium, and tritium.

20. All three isotopes of hydrogen have the same number of protons (1) and electrons (1). They differ in the number of neutrons (0,1,and 2).

21. $^{52}_{24}$Cr   chromium – 52

22. (a) $201 - 121 = 80$ protons; electrical charge of the nucleus is $+80$

    (b)   Hg, mercury

23. All six isotopes have 20 protons and 20 electrons. The number of neutrons are

Isotope	Neutrons
40	20
42	22
43	23
44	24
46	26
48	28

24. The presence of isotopes indicates that atomic mass is not as unique a characteristric of an element as is atomic number. The atomic mass of lanthanum is the average mass of the four isotopes. It primarily reflects the mass of the most abundant isotope.

25. An atom is electrically neutral containing equal numbers of electrons and protons. An ion is charged, containing more electrons than protons if it is an anion (negative) and less electrons than protons if it is a cation (positive).

26. 42 protons $+54$ neutrons $= 96$ nucleons, each with a mass of 1 amu. The mass of electrons is negligible. The approximate atomic mass of the element is 96 amu.

27. Atoms of an element having the same atomic number but differing atomic masses are isotopes. Examples include:

    $^{12}_{6}$C, $^{14}_{6}$C; $^{238}_{92}$U, $^{235}_{92}$U; $^{1}_{1}$H, $^{2}_{1}$H, $^{3}_{1}$H; $^{16}_{8}$O, $^{17}_{8}$O, $^{18}_{8}$O.

28. The correct statements are a, b, d.

    (c) $^{23}_{11}$Na and $^{24}_{11}$Na have different atomic masses.

29. A wintergreen lifesaver is an example of a substance which exihibits triboluminescence. The wintergreen molecules absorb some of the energy and re-emit it as blu-green light.

30. Synthetic and natural vanilla contain exactly the same molecules. The difference between them is in the source of the molecules of vanillin. Natural vanilla comes from the vanilla bean, while the synthetic variety is made from lignin.

31. The SIRA technique is stable isotope ratio analysis. The carbon atoms in a molecule are analyzed to determine the ratio of C-12 and C-13. The ratio is differnet for synthetic and natural vanillin and so can be used to differentiate between these types.

32.

	protons	neutrons	electrons
He	2	2	2
C	6	6	6
N	7	7	7
O	8	8	8
Ne	10	10	10
Mg	12	12	12
Si	14	14	14
S	16	16	16
Ca	20	20	20

33.

	Atomic number	Mass number	Symbol	Protons	Neutrons
(a)	8	16	O	8	8
(b)	28	58	Ni	28	30
(c)	80	199	Hg	80	119

34. $(2.18 \times 10^{-22} \text{ amu})\left(\dfrac{12.0 \text{ amu C-12}}{1.9927 \times 10^{-23} \text{ amu (C)}}\right) = 131 \text{ amu}$

35. $(0.5182)(106.9041 \text{ amu}) + (0.4818)(108.9047 \text{ amu}) = x$
$\qquad 55.40 \text{ amu} + 52.47 \text{ amu} = x$
$\qquad 107.9 \text{ amu} = \text{atomic mass Ag}$

36. $(0.7899)(23.985 \text{ amu}) + (0.1000)(24.986 \text{ amu}) + (0.1101)(25.983 \text{ amu})$
$= 18.95 \text{ amu} + 2.500 \text{ amu} + 2.861 \text{ amu}$
$= 24.31 \text{ amu} = \text{atomic mass Mg}$

37. $(40.0 \text{ g})\left(\dfrac{1 \text{ atom}}{6.63 \times 10^{-24} \text{ g}}\right) = 6.03 \times 10^{24} \text{ atoms}$

38. $(0.604)(68.9257 \text{ amu}) + (1.00 - 0.604)(70.9249 \text{ amu})$
$= 41.6 \text{ amu} + 28.1 \text{ amu} = 69.7 \text{ amu} = \text{atomic mass}$
The element is gallium.

39.

	Element	Symbol	Atomic #	Protons	Neutrons	Electrons
(a)	platinum	^{195}Pt	78	78	117	78
(b)	phosphorus	^{30}P	15	15	15	15
(c)	iodine	^{127}I	53	53	74	53
(d)	krypton	^{84}Kr	36	36	48	36
(e)	selenium	^{79}Se	34	34	45	34
(f)	calcium	^{40}Ca	20	20	20	20

NOMENCLATURE OF INORGANIC COMPOUNDS

1. (a) CaO, NiO, Ag_2O, Al_2O_3, Na_2O

   (b) LiI, AlI_3, BI_3, MgI_2

2. (a) $NaClO_3$ (d) Cu_2O

   (b) H_2SO_4 (e) $Zn(HCO_3)_2$

   (c) $Sn(C_2H_3O_2)_2$ (f) $Fe_2(CO_3)_3$

3. Oxidation numbers of underlined elements.

   (a) $\underline{Mn}CO_3$ +2   (e) $Ba\underline{C}O_3$ +4

   (b) $\underline{Sn}F_4$ +4   (f) $\underline{P}Cl_3$ +3

   (c) $K\underline{N}O_3$ +5   (g) $\underline{W}Cl_5$ +5

   (d) $K\underline{Mn}O_4$ +7   (h) $K_2\underline{Cr}_2O_7$ +6

4. Oxidation numbers of underlined elements.

   (a) $\underline{In}I_3$ +3   (e) $Mg(\underline{N}O_3)_2$ +5

   (b) $K\underline{Cl}O_3$ +5   (f) $\underline{Sn}O_2$ +4

   (c) $Na_2\underline{S}O_4$ +6   (g) $\underline{Fe}_2(CO_3)_3$ +3

   (d) $\underline{C}_2H_5OH$ -2   (h) $Na\underline{Cl}O_4$ +7

5. Formulas of compounds.

   (a) Na and I: NaI (d) K and S: K_2S

   (b) Ba and F: BaF_2 (e) Cs and Cl: $CsCl$

   (c) Al and O: Al_2O_3 (f) Sr and Br: $SrBr_2$

6. Formulas of compounds.

   (a) Ba and O: BaO (d) Be and Br: $BeBr_2$

   (b) H and S: H_2S (e) Li and Si: Li_4Si

   (c) Al and Cl: $AlCl_3$ (f) Mg and P: Mg_3P_2

7. No. It means their oxidation numbers, while of opposite sign, are of the same magnitude.

8.

sodium	Na^+	cobalt(II)	Co^{2+}
magnesium	Mg^{2+}	hydrogen	H^+
aluminum	Al^{3+}	mercury(II)	Hg^{2+}
copper(II)	Cu^{2+}	tin(II)	Sn^{2+}
barium	Ba^{2+}	chromium(III)	Cr^{3+}
iron(II)	Fe^{2+}	stannic	Sn^{4+}
ferric	Fe^{3+}	manganese(II)	Mn^{2+}
lead(II)	Pb^{2+}	bismuth(III)	Bi^{3+}
silver	Ag^+		

9.

chloride	Cl^-	bisulfate	HSO_4^-
bromide	Br^-	bisulfite	HSO_3^-
fluoride	F^-	chromate	CrO_4^{2-}
iodide	I^-	carbonate	CO_3^{2-}
cyanide	CN^-	bicarbonate	HCO_3^-
oxide	O^{2-}	acetate	$C_2H_3O_2^-$
hydroxide	OH^-	chlorate	ClO_3^-
sulfide	S^{2-}	permanganate	MnO_4^-
sulfate	SO_4^{2-}	oxalate	$C_2O_4^{2-}$

10.

Ion	Br^-	O^{2-}	NO_3^-	PO_4^{3-}	CO_3^{2-}
K^+	KBr	K_2O	KNO_3	K_3PO_4	K_2CO_3
Mg^{2+}	$MgBr_2$	MgO	$Mg(NO_3)_2$	$Mg_3(PO_4)_2$	$MgCO_3$
Al^{3+}	$AlBr_3$	Al_2O_3	$Al(NO_3)_3$	$AlPO_4$	$Al_2(CO_3)_3$
Zn^{2+}	$ZnBr_2$	ZnO	$Zn(NO_3)_2$	$Zn_3(PO_4)_2$	$ZnCO_3$
H^+	HBr	H_2O	HNO_3	H_3PO_4	H_2CO_3

11.

Ion	SO_4^{2-}	Cl^-	AsO_4^{3-}	$C_2H_3O_2^-$	CrO_4^{2-}
NH_4^+	$(NH_4)_2SO_4$	NH_4Cl	$(NH_4)_3AsO_4$	$NH_4C_2H_3O_2$	$(NH_4)_2CrO_4$
Ca^{2+}	$CaSO_4$	$CaCl_2$	$Ca_3(AsO_4)_2$	$Ca(C_2H_3O_2)_2$	$CaCrO_4$
Fe^{3+}	$Fe_2(SO_4)_3$	$FeCl_3$	$FeAsO_4$	$Fe(C_2H_3O_2)_3$	$Fe_2(CrO_4)_3$
Ag^+	Ag_2SO_4	$AgCl$	Ag_3AsO_4	$AgC_2H_3O_2$	Ag_2CrO_4
Cu^{2+}	$CuSO_4$	$CuCl_2$	$Cu_3(AsO_4)_2$	$Cu(C_2H_3O_2)_2$	$CuCrO_4$

12. ide: suffix is used to indicate a binary compound except for hydroxides, cyanides, and ammonium compounds.

ous: used in acids to indicate a lower oxidation state of any element other than hydrogen and oxygen; also used for the lower oxidation state of a multivalent metal.

hypo: used as a prefix in acids or salts when the oxidation state is lower than that of -ous acid or the -ite salt.

per: used as a prefix in acids or salts when the oxidation state is greater than that of the -ic acid or the -ate salt.

ite: the suffix of a salt derived from an -ous acid.

ate: the suffix of a salt derived from an -ic acid.

Roman numerals: In the Stock System Roman numerals are used in naming compounds that contain metals that may exist in more than one oxidation state. The oxidation number of a metal is indicated by a Roman numeral written in parentheses immediately after the name of the metal.

13. Nonmetal binary compound formulas:

(a) carbon monoxide, CO

(b) sulfur trioxide, SO_3

(c) carbon tetrabromide, CBr_4

(d) phosphorus trichloride, PCl_3

(e) nitrogen dioxide, NO_2

(f) dinitrogen pentoxide, N_2O_5

14. Naming binary nonmetal compounds:

(a) CO_2 carbon dioxide

(b) N_2O dinitrogen oxide

(c) PCl_5 phosphorus pentachloride

(d) CCl_4 carbon tetrachloride

(e) SO_2 sulfur dioxide

(f) N_2O_4 dinitrogen tetraoxide

(g) P_2O_5 diphosphorus pentoxide

(h) OF_2 oxygen difluoride

(i) NF_3 nitrogen trifluoride

(j) CS_2 carbon disulfide

15. (a) K_2O, potassium oxide

(b) NH_4Br, ammonium bromide

(c) CaI_2, calcium iodide

(d) $BaCO_3$, barium carbonate

(e) Na_3PO_4, sodium phosphate

(f) Al_2O_3, aluminum oxide

(g) $Zn(NO_3)_2$, zinc nitrate

(h) Ag_2SO_4, silver sulfate

16. (a) sodium nitrate, $NaNO_3$ (d) ammonium sulfate, $(NH_4)_2SO_4$

    (b) magnesium fluoride, MgF_2 (e) silver carbonate, Ag_2CO_3

    (c) barium hydroxide, $Ba(OH)_2$ (f) calcium phosphate, $Ca_3(PO_4)_2$

17. (a) $CuCl_2$ copper(II) chloride   cupric chloride

    (b) $CuBr$ copper(I) bromide   cuprous bromide

    (c) $Fe(NO_3)_2$ iron(II) nitrate   ferrous nitrate

    (d) $FeCl_3$ iron(III) chloride   ferric chloride

    (e) SnF_2 tin(II) fluoride   stannous fluoride

    (f) $HgCO_3$ mercury(II) carbonate   mercuric carbonate

    (g) $As(C_2H_3O_2)_3$ arsenic(III) acetate   arsenous acetate

    (h) TiI_3 titanium(III) iodide   titanous iodide

18. Formulas:

    (a) tin(IV) bromide $SnBr_4$

    (b) copper(I) sulfate Cu_2SO_4

    (c) ferric carbonate $Fe_2(CO_3)_3$

    (d) mercuric nitrite $Hg(NO_2)_2$

    (e) titanic sulfide TiS_2

    (f) iron(II) acetate $Fe(C_2H_3O_2)_2$

19. Acid formulas:

    (a) hydrochloric acid, HCl (d) carbonic acid, H_2CO_3

    (b) chloric acid, $HClO_3$ (e) sulfurous acid, H_2SO_3

    (c) nitric acid, HNO_3 (f) phosphoric acid, H_3PO_4

20. Naming acids:

    (a) HNO_2, nitrous acid   (e) H_3PO_3, phosphorous acid

    (b) H_2SO_4, sulfuric acid   (f) $HC_2H_3O_2$, acetic acid

    (c) $H_2C_2O_4$, oxalic acid   (g) HF, hydrofluoric acid

    (d) HBr, hydrobromic acid   (h) $HBrO_3$, bromic acid

21. Formulas of acids:

    (a) acetic acid, $HC_2H_3O_2$ (d) boric acid, H_3BO_3

    (b) hydrofluoric acid, HF (e) nitrous acid, HNO_2

    (c) hypochlorous acid, $HClO$ (f) hydrosulfuric acid, H_2S

22. Naming acids:

   (a) H_3PO_4, phosphoric acid

   (b) H_2CO_3, carbonic acid

   (c) HIO_3, iodic acid

   (d)   HCl, hydrochloric acid

   (e)   HClO, hypochlorous acid

   (f) HNO_3, nitric acid

   (g)   HI, hydroiodic acid

   (h) $HClO_4$, perchloric acid

23. Naming compounds:

   (a) $Ba(NO_3)_2$, barium nitrate

   (b) $NaC_2H_3O_2$, sodium acetate

   (c) PbI_2, lead(II) iodide

   (d) $MgSO_4$, magnesium sulfate

   (e) $CdCrO_4$ cadmium chromate

   (f) $BiCl_3$ bismuth(III) chloride

   (g)   NiS, nickel(II) sulfide

   (h) $Sn(NO_3)_4$, tin(IV) nitrate

   (i) $Ca(OH)_2$, calcium hydroxide

24. Formulas for:

   (a)   silver sulfite Ag_2SO_3

   (b)   cobalt(II) bromide $CoBr_2$

   (c)   tin(II) hydroxide $Sn(OH)_2$

   (d)   aluminum sulfate $Al_2(SO_4)_3$

   (e)   manganese(II) fluoride MnF_2

   (f)   ammonium carbonate $(NH_4)_2CO_3$

   (g)   chromium(III) oxide Cr_2O_3

   (h)   cupric chloride $CuCl_2$

   (i)   potassium permanganate $KMnO_4$

   (j)   barium nitrite $Ba(NO_2)_2$

   (k)   sodium peroxide Na_2O_2

   (l)   ferrous sulfate $FeSO_4$

   (m)   potassium dichromate $K_2Cr_2O_7$

   (n)   bismuth(III) chromate $Bi_2(CrO_4)_3$

25. Formulas for:

   (a)   sodium chromate Na_2CrO_4

   (b)   magnesium hydride MgH_2

   (c)   nickel(II) acetate $Ni(C_2H_3O_2)_2$

(d)   calcium chlorate $Ca(ClO_3)_2$

(e)   lead(II) nitrate $Pb(NO_3)_2$

(f)   potassium dihydrogen KH_2PO_4
      phosphate

(g)   manganese(II) hydroxide $Mn(OH)_2$

(h)   cobalt(II) bicarbonate $Co(HCO_3)_2$

(i)   sodium hypochlorite $NaClO$

(j)   arsenic(V) carbonate $As_2(CO_3)_5$

(k)   chromium(III) sulfite $Cr_2(SO_3)_3$

(l)   antimony(III) sulfate $Sb_2(SO_4)_3$

(m)   sodium oxalate $Na_2C_2O_4$

(n)   potassium thiocyanate $KSCN$

26.

	Salt Formula	Salt Name	Acid Formula	Acid Name
(a)	$ZnSO_4$	zinc sulfate	H_2SO_4	sulfuric acid
(b)	$HgCl_2$	mercury(II) choride	HCl	hydrochloric acid
(c)	$CuCO_3$	copper(II) carbonate	H_2CO_3	carbonic acid
(d)	$Cd(NO_3)_2$	cadmium nitrate	HNO_3	nitric acid
(e)	$Al(C_2H_3O_2)_3$	aluminum acetate	$HC_2H_3O_2$	acetic acid
(f)	CoF_2	cobalt(II) fluoride	HF	hydrofluoric acid
(g)	$Cr(ClO_3)_3$	chromium(III) chlorate	$HClO_3$	chloric acid
(h)	Ag_3PO_4	silver phosphate	H_3PO_4	phosphoric acid
(i)	NiS	nickel(II) sulfide	H_2S	hydrosulfuric acid
(j)	$BaCrO_4$	barium chromate	H_2CrO_4	chromic acid
(k)	$Ca(HSO_4)_2$	calcium bisulfate	H_2SO_4	sulfuric acid
(l)	$As_2(SO_3)_3$	arsenic(III) sulfite	H_2SO_3	sulfurous acid
(m)	$Sn(NO_2)_2$	tin(II) nitrite	HNO_2	nitrous acid
(n)	$FeBr_3$	iron(III) bromide	HBr	hydrobromic acid
(o)	$KHCO_3$	potassium bicarbonate	H_2CO_3	carbonic acid
(p)	$BiAsO_4$	bismuth(III) arsenate	H_3AsO_4	arsenic acid
(q)	$Fe(BrO_3)_2$	iron(II) bromate	$HBrO_3$	bromic acid
(r)	$(NH_4)_2HPO_4$	ammonium mono- hydrogen phosphate	H_3PO_4	phosphoric acid
(s)	$NaClO$	sodium hypochlorite	$HClO$	hypochorous acid
(t)	$KMnO_4$	potassium permanganate	$HMnO_4$	permanganic acid

27. Formulas for:

   (a)   baking soda $NaHCO_3$

   (b)   lime CaO

   (c)   Epsom salts $MgSO_4 \cdot 7\,H_2O$

   (d)   muriatic acid HCl

   (e)   vinegar $HC_2H_3O_2$

   (f)   potash K_2CO_3

   (g)   lye $NaOH$

   (h)   quicksilver Hg

   (i)   fool's gold FeS_2

   (j)   saltpeter $NaNO_3$

   (k)   limestone $CaCO_3$

   (l)   cane sugar $C_{12}H_{22}O_{11}$

   (m)   milk of magnesia $Mg(OH)_2$

   (n)   washing soda $Na_2CO_3 \cdot 10\,H_2O$

   (o)   grain alcohol C_2H_5OH

28. Hard water contains calcium and magnesium ions which form precipitates in the presence of soap. These precipitates stick to the clothing fiber dulling the appearance of white laundry.

29. Phosphates are added to detergents for two reasons: 1) to improve the function of cleaning agents by changing the acidity of the wash solution and 2) to prevent the formation of insoluble precipitates.

30.   calcium cyanide $Ca(CN)_2$

           ammonium bromide NH_4Br

           lithium cyanide $LiCN$

31. The correct statements are: a, d, f, g, i, j, l, m, o, p, r, u, w, x, y, aa.

   (b)   The oxidation number of Li, Na, and K compounds is +1.

   (c)   The sum of the oxidation numbers for all the atoms in a polyatomic ion is equal to the charge on the ion.

   (e)   All the following compounds are acids: H_2SO_4, HCl, HNO_3.

   (h)   The formula for the compound between Fe^{3+} and O^{2-} is Fe_2O_3.

   (k)   The oxidation number of Co in $CoCl_2$ is +2.

   (n)   The name for CuO is copper(II) oxide.

   (q)   The name for Na_2O is sodium oxide.

(s)   If the name of an anion ends with <u>ite</u>, the corresponding acid name will end in <u>ous</u>.

(t)   If the name of an acid ends in <u>ous</u>, the corresponding salt will end with <u>ite</u>.

(v)   In Cu_2SO_4, the copper ion is copper(I) because two copper ions are combined with a -2 sulfate ion.

(z) $Sn(CrO_4)_2$ is called tin(IV) chromate.

QUANTITATIVE COMPOSITION OF COMPOUNDS

1. A mole is an amount of substance containing the same number of formula units as there are atoms in exactly 12 g of carbon–12.

    It is Avogadro's number (6.022×10^{23}) of anything (atoms, molecules, ping-pong balls, etc).

2. A mole of gold has a higher mass than a mole of potassium.

3. Both samples (Au and K) contain the same number of atoms.   (6.022×10^{23}).

4. A mole of gold atoms contains more electrons than a mole of potassium atoms, as each Au atom has 79 e^-, while each K atom has only 19 e^-.

5. No.   Avogadro's number is a constant.   The mole is defined as Avogadro's number of C–12 atoms.   Changing the atomic mass to 50 amu would change only the size of the atomic mass unit, not Avogadro's number.

6. 6.022×10^{23}

7. There are Avogadro's number of particles in one mole of substance.

8. (a)   A mole of oxygen atoms (O) contains **6.022×10^{23}** atoms.

    (b)   A mole of oxygen molecules (O_2) contains **6.022×10^{23}** molecules.

    (c)   A mole of oxygen molecules (O_2) contains **1.204×10^{24}** atoms.

    (d)   A mole of oxygen atoms (O) has a mass of **16.00 g.**

    (e)   A mole of oxygen molecules (O_2) has a mass of **32.00 g.**

9. The correct answers are a, b, c, d, h.

    (e)   A mole of chlorine molecules (Cl_2) contains 1.204×10^{24} atoms of chlorine.

    (f)   A mole of aluminum atoms has a different mass than a mole of tin atoms.

    (g)   A mole of H_2O contains 6.022×10^{23} molecules.

10. 6.022×10^{23} molecules in one molar mass of H_2SO_4.

    4.215×10^{24} atoms in one molar mass of H_2SO_4.

11. Choosing 100 g of a compound allows us to simply drop the % sign and use grams for each percent.

12. Food additives are generally preservatives, coloring, flavorings, antioxidants or sweeteners. They are used to enhance the flavor or appearance of food or to help preserve it for later use.

13. The maximum daily tolerable intake of an additive is determined by calculation based on body mass in mg additive/kg mass/day.

14. It is difficult to determine safe levels of food additives because testing is done on animals which are not always the same as humans. Variation also exists between species regarding sensitivity for an additive. Children with different metabolic rates also pose a problem in determining safe levels.

15. The correct answers are a, d, f, g, j, k, m.

   (b) One mole of nitrogen gas (N_2) has a mass of 28.02 g.

   (c) The percent of oxygen is higher in Na_2CrO_4 than it is in K_2CrO_4.

   (e) K_2CrO_4 and Na_2CrO_4 have different percentages of Cr by mass.

   (h) The empirical formula for sucrose is $C_{12}H_{22}O_{11}$.

   (i) A hydrocarbon that has a molar mass of 280 and an empirical formula of CH_2 has a molecular formula of $C_{20}H_{40}$.

   (l) If the molecular formula and the empirical formula of a compound are not the same, the molecular formula will be an integral multiple of the empirical formula.

   (n) A compound having an empirical formula of CH_2O, and a molar mass of 60, has a molecular formula of $C_2H_4O_2$.

16. Molar masses

   (a) KBr

1	K	39.10
1	Br	79.90
		119.0

   (b) Na_2SO_4

2	Na	45.98
1	S	32.06
4	O	64.00
		142.0

   (c) $Pb(NO_3)_2$

1	Pb	207.2
2	N	28.02
6	O	96.00
		331.2

   (d) C_2H_5OH

2	C	24.02
6	H	6.048
1	O	16.00
		46.07

   (e) $HC_2H_3O_2$

4	H	4.032
2	C	24.02
2	O	32.00
		60.05

(f)	Fe_3O_4	3	Fe	167.6	
		4	O	64.00	
				231.6	

(g)	$C_{12}H_{22}O_{11}$	12	C	144.1	
		22	H	22.18	
		11	O	176.0	
				342.3	

(h)	$Al_2(SO_4)_3$	2	Al	53.96	
		3	S	96.18	
		12	O	192.0	
				342.1	

(i)	$(NH_4)_2HPO_4$	9	H	9.072	
		2	N	28.02	
		1	P	30.97	
		4	O	64.00	
				132.1	

17. Molar masses

(a)	NaOH	1	Na	22.99	
		1	O	16.00	
		1	H	1.008	
				40.00	

(b)	Ag_2CO_3	2	Ag	215.8	
		1	C	12.01	
		3	O	48.00	
				275.8	

(c)	Cr_2O_3	2	Cr	104.0	
		3	O	48.00	
				152.0	

(d)	$(NH_4)_2CO_3$	2	N	28.02	
		8	H	8.064	
		1	C	12.01	
		3	O	48.00	
				96.09	

(e)	$Mg(HCO_3)_2$	1	Mg	24.31	
		2	H	2.016	
		2	C	24.02	
		6	O	96.00	
				146.3	

(f)	C_6H_5COOH	7	C	84.07	
		6	H	6.048	
		2	O	32.00	
				122.1	

(g) $C_6H_{12}O_6$

6	C	72.06
12	H	12.10
6	O	96.00
		180.2

(h) $K_4Fe(CN)_6$

4	K	156.4
1	Fe	55.85
6	C	72.06
6	N	84.06
		368.4

(i) $BaCl_2 \cdot 2\ H_2O$

1	Ba	137.3
2	Cl	70.90
4	H	4.072
2	O	32.00
		244.2

18. Moles of atoms.

(a) $(22.5 \text{ g Zn})\left(\dfrac{1 \text{ mol}}{65.38 \text{ g}}\right) = 0.344 \text{ mol Zn}$

(b) $(0.688 \text{ g Mg})\left(\dfrac{1 \text{ mol}}{24.31 \text{ g}}\right) = 2.83 \times 10^{-2} \text{ mol Mg}$

(c) $(4.5 \times 10^{22} \text{ atoms Cu})\left(\dfrac{1 \text{ mol}}{6.022 \times 10^{23} \text{atoms}}\right) = 7.5 \times 10^{-2} \text{ mol Cu}$

(d) $(382 \text{ g Co})\left(\dfrac{1 \text{ mol}}{58.93 \text{ g}}\right) = 6.48 \text{ mol Co}$

(e) $0.055 \text{ g Sn})\left(\dfrac{1 \text{ mol}}{118.7 \text{ g}}\right) = 4.6 \times 10^{-4} \text{ mol Sn}$

(f) $(8.4 \times 10^{24} \text{ molecules N}_2)\left(\dfrac{2 \text{ atoms N}}{1 \text{ molecule N}_2}\right)\left(\dfrac{1 \text{ mol N atoms}}{6.022\ 10^{23} \text{ atoms N}}\right) = 28 \text{ mol N atoms}$

19. Number of moles.

(a) $(25.0 \text{ g NaOH})\left(\dfrac{1 \text{ mol}}{40.00 \text{ g}}\right) = 0.625 \text{ mol NaOH}$

(b) $(44.0 \text{ g Br}_2)\left(\dfrac{1 \text{ mol}}{159.8 \text{ g}}\right) = 0.275 \text{ mol Br}_2$

(c) $(0.684 \text{ g MgCl}_2)\left(\dfrac{1 \text{ mol}}{95.21 \text{ g}}\right) = 7.18 \times 10^{-3} \text{ mol MgCl}_2$

(d) $(14.8 \text{ g CH}_3\text{OH})\left(\dfrac{1 \text{ mol}}{32.04 \text{ g}}\right) = 0.462 \text{ mol CH}_3\text{OH}$

(e) $(2.88 \text{ g Na}_2\text{SO}_4)\left(\dfrac{1 \text{ mol}}{142.0 \text{ g}}\right) = 2.03 \times 10^{-2} \text{ mol Na}_2\text{SO}_4$

(f) $(4.20 \text{ lb ZnI}_2)\left(\dfrac{454 \text{ g}}{1 \text{ lb}}\right)\left(\dfrac{1\text{mol}}{319.2 \text{ g}}\right) = 5.97 \text{ mol ZnI}_2$

20. Number of grams.

(a) $(0.550 \text{ mol Au})\left(\dfrac{197.0 \text{ g}}{1 \text{ mol}}\right) = 108 \text{ g Au}$

(b) $(15.8 \text{ mol H}_2\text{O})\left(\dfrac{18.02 \text{ g}}{\text{mol}}\right) = 285 \text{ g H}_2\text{O}$

(c) $(12.5 \text{ mol Cl}_2)\left(\dfrac{70.90 \text{ g}}{\text{mol}}\right) = 886 \text{ g Cl}_2$

(d) $(3.15 \text{ mol NH}_4\text{NO}_3)\left(\dfrac{80.10 \text{ g}}{\text{mol}}\right) = 252 \text{ g NH}_4\text{NO}_3$

(e) $(4.25 \times 10^{-4} \text{ mol H}_2\text{SO}_4)\left(\dfrac{98.08 \text{ g}}{\text{mol}}\right) = 0.0417 \text{ g H}_2\text{SO}_4$

(f) $(4.5 \times 10^{22} \text{ molecules CCl}_4)\left(\dfrac{1 \text{ mol}}{6.022 \times 10^{23} \text{molecules}}\right)\left(\dfrac{153.8 \text{ g}}{\text{mol}}\right) = 11 \text{ g CCl}_4$

(g) $(0.00255 \text{ mol Ti})\left(\dfrac{47.90 \text{ g}}{\text{mol}}\right) = 0.122 \text{ g Ti}$

(h) $(1.5 \times 10^{16} \text{atoms S})\left(\dfrac{32.06 \text{ g}}{6.022 \times 10^{23} \text{ atoms}}\right) = 8.0 \times 10^{-7} \text{ g S}$

21. Number of molecules.

(a) $(1.26 \text{ mol O}_2)\left(\dfrac{6.022 \times 10^{23} \text{ molecules}}{\text{mol}}\right) = 7.59 \times 10^{23} \text{molecules O}_2$

(b) $(0.56 \text{ mol C}_6\text{H}_6)\left(\dfrac{6.022 \times 10^{23} \text{ molecules}}{\text{mol}}\right) = 3.4 \times 10^{23} \text{ molecules C}_6\text{H}_6$

(c) $(16.0 \text{ g CH}_4)\left(\dfrac{6.022 \times 10^{23} \text{ molecules}}{16.04 \text{ g}}\right) = 6.022 \times 10^{23} \text{ molecules CH}_4$

(d) $(1000. \text{ g HCl})\left(\dfrac{6.022 \times 10^{23} \text{ molecules}}{36.46 \text{ g}}\right) = 1.652 \times 10^{25} \text{ molecules HCl}$

22. Number of grams.

(a) $(1 \text{ atom Pb})\left(\dfrac{207.2 \text{ g}}{6.022 \times 10^{23} \text{ atoms}}\right) = 3.441 \times 10^{-22} \text{ g Pb}$

(b) $(1 \text{ atom Ag})\left(\dfrac{107.9 \text{ g}}{6.022 \times 10^{23} \text{ atoms}}\right) = 1.791 \times 10^{-22} \text{ g Ag}$

(c) $(1 \text{ molecule H}_2\text{O})\left(\dfrac{18.02 \text{ g}}{6.022 \times 10^{23} \text{ molecules}}\right) = 2.992 \times 10^{-23} \text{ g H}_2\text{O}$

(d) $(1 \text{ molecule C}_3\text{H}_5(\text{NO}_3)_3)\left(\dfrac{227.1 \text{ g}}{6.022 \times 10^{23} \text{ molecules}}\right) = 3.771 \times 10^{-22} \text{ g C}_3\text{H}_5(\text{NO}_3)_3$

23. (a) $(8.66 \text{ mol Cu})\left(\dfrac{63.55 \text{ g}}{\text{mol}}\right) = 550. \text{ g Cu}$

(b) $(125 \text{ mol Au})\left(\dfrac{197.0 \text{ g}}{\text{mol}}\right)\left(\dfrac{1 \text{ kg}}{1000 \text{ g}}\right) = 24.6 \text{ kg Au}$

(c) $(10 \text{ atoms C})\left(\dfrac{1 \text{ mol}}{6.022 \times 10^{23} \text{ atoms}}\right) = 1.7 \times 10^{-23} \text{ mol C}$

(d) $(5000 \text{ molecules CO}_2)\left(\dfrac{1 \text{ mol}}{6.022 \times 10^{23} \text{ molecules}}\right) = 8.3 \times 10^{-21} \text{ mol CO}_2$

(e) $(28.4 \text{ g S})\left(\dfrac{1 \text{ mol}}{32.06 \text{ g}}\right) = 0.886 \text{ mol S}$

(f) $(2.50 \text{ kg NaCl})\left(\dfrac{1000 \text{ g}}{\text{kg}}\right)\left(\dfrac{1 \text{ mol}}{58.44 \text{ g}}\right) = 42.8 \text{ mol NaCl}$

(g) $(42.4 \text{ g Mg})\left(\dfrac{6.022 \times 10^{23} \text{ atoms}}{24.31 \text{ g}}\right) = 1.05 \times 10^{24} \text{ atoms Mg}$

(h) $(485 \text{ mL Br}_2)\left(\dfrac{3.12 \text{ g}}{\text{mL}}\right)\left(\dfrac{1 \text{ mol}}{159.8 \text{ g}}\right) = 9.47 \text{ mol Br}_2$

24. One mole of carbon disulfide (CS_2) contains:

(a) 6.022×10^{23} molecules of CS_2

(b) $(6.022 \times 10^{23} \text{ molecules CS}_2)\left(\dfrac{1 \text{ C atom}}{1 \text{ molecule CS}_2}\right) = 6.022 \times 10^{23} \text{ C atoms}$

(c) $(6.022 \times 10^{23} \text{ molecules CS}_2)\left(\dfrac{2 \text{ S atoms}}{1 \text{ molecule CS}_2}\right) = 1.204 \times 10^{24} \text{ S atoms}$

(d) $6.022 \times 10^{23} \text{ atoms} + 1.204 \times 10^{24} \text{ atoms} = 1.806 \times 10^{24} \text{ atoms}$

25. $(0.350 \text{ mol P}_4)\left(\dfrac{6.022 \times 10^{23} \text{ molecules}}{\text{mol}}\right)\left(\dfrac{4 \text{ atoms P}}{\text{molecule P}_4}\right) = 8.43 \times 10^{23} \text{ atoms P}$

26. $(10.0 \text{ g K})\left(\dfrac{1 \text{ mol}}{39.10 \text{ g}}\right)\left(\dfrac{1 \text{ mol Na}}{1 \text{ mol K}}\right)\left(\dfrac{22.99 \text{ g}}{\text{mol}}\right) = 5.88 \text{ g Na}$

27. $(1.79 \times 10^{-23} \text{ g/atom})(6.022 \times 10^{23} \text{ atoms/atomic mass}) = 10.8 \text{ g/atomic mass}$

28. $(6.022 \times 10^{23} \text{ sheets})\left(\dfrac{4.60 \text{ cm}}{500 \text{ sheets}}\right)\left(\dfrac{1 \text{ m}}{100 \text{ cm}}\right) = 5.54 \times 10^{19} \text{ m}$

29. $\left(\dfrac{6.022 \times 10^{23} \text{ dollars}}{5.0 \times 10^9 \text{ people}}\right) = 1.2 \times 10^{14} \text{ dollars/person}$

30. (a) $(1 \text{ mi}^3)\left(\dfrac{5280 \text{ ft}}{\text{mile}}\right)^3\left(\dfrac{12.0 \text{ in.}}{\text{ft}}\right)^3\left(\dfrac{2.54 \text{ cm}}{\text{inch}}\right)^3\left(\dfrac{20 \text{ drops}}{1.0 \text{ cm}^3}\right) = 8.3 \times 10^{16} \text{ drops}$

(b) $(6.022 \times 10^{23} \text{ drops})\left(\dfrac{1 \text{ mi}^3}{8.3 \times 10^{16} \text{ drops}}\right) = 7.3 \times 10^6 \text{ mi}^3$

31.   1 mol Ag $=$ 107.9 g

(a) $(107.9 \text{ g Ag})\left(\dfrac{1 \text{ cm}^3}{10.5 \text{ g}}\right) = 10.3 \text{ cm}^3$

(b) $10.3 \text{ cm}^3 = \text{volume of cube} = (\text{side})^3$

     $\text{side} = \sqrt[3]{\dfrac{10.3 \text{ cm}^3}{1}} = 2.18 \text{ cm}$

32. Atoms of oxygen in:

(a) $(16.0 \text{ g O}_2)\left(\dfrac{1 \text{ mol}}{16.00 \text{ g}}\right)\left(\dfrac{6.022 \times 10^{23} \text{ atoms}}{\text{mol}}\right) = 6.02 \times 10^{23} \text{ atoms O}$

(b) $(0.622 \text{ mol MgO})\left(\dfrac{1 \text{ mol O}}{\text{mol MgO}}\right)\left(\dfrac{6.022 \times 10^{23} \text{ atoms}}{\text{mol O}}\right) = 3.75 \times 10^{23} \text{ atoms O}$

(c) $(6.00 \times 10^{22} \text{ molecules C}_6\text{H}_{12}\text{O}_6)\left(\dfrac{6 \text{ atoms O}}{\text{molecule C}_6\text{H}_{12}\text{O}_6}\right) = 3.60 \times 10^{23} \text{ atoms O}$

(d) $(5.0 \text{ mol MnO}_2)\left(\dfrac{2 \text{ mol O}}{\text{mol MnO}_2}\right)\left(\dfrac{6.022 \times 10^{23} \text{ atoms}}{\text{mol}}\right) = 6.0 \times 10^{24} \text{ atoms O}$

(e) $(250 \text{ g MgCO}_3)\left(\dfrac{1 \text{ mol}}{84.32 \text{ g}}\right)\left(\dfrac{3 \text{ mol O}}{\text{mol MgCO}_3}\right)\left(\dfrac{6.022 \times 10^{23} \text{ atoms O}}{\text{mol}}\right) = 5.4 \times 10^{24} \text{ atoms O}$

(f) $(5.0 \times 10^{18} \text{ molecules H}_2\text{O})\left(\dfrac{1 \text{ atom O}}{\text{molecule H}_2\text{O}}\right) = 5.0 \times 10^{18} \text{ atoms O}$

33. The number of grams of:

(a) silver in 25.0 g AgBr

     $(25.0 \text{ g AgBr})\left(\dfrac{107.9 \text{ g Ag}}{187.8 \text{ g AgBr}}\right) = 14.4 \text{ g Ag}$

(b) chlorine in 5.00 g $PbCl_2$

     $(5.00 \text{ g PbCl}_2)\left(\dfrac{70.90 \text{ g Cl}}{278.1 \text{ g PbCl}_2}\right) = 1.27 \text{ g Cl}$

(c) nitrogen in 6.34 mol $(NH_4)_3PO_4$

     $(6.34 \text{ mol (NH}_4)_3\text{PO}_4)\left(\dfrac{42.03 \text{ g N}}{\text{mol (NH}_4)_3\text{PO}_4}\right) = 266 \text{ g N}$

(d) oxygen in 8.45×10^{22} molecules SO_3

     $(8.45 \times 10^{22} \text{ molecules SO}_3)\left(\dfrac{1 \text{ mol}}{6.022 \times 10^{23} \text{ molecules}}\right)\left(\dfrac{48.00 \text{ g O}}{\text{mol SO}_3}\right) = 6.74 \text{ g O}$

(e) hydrogen in 45.0 g C_3H_8O

$$(45.0 \text{ g } C_3H_8O)\left(\frac{8.064 \text{ g H}}{60.09 \text{ g } C_3H_8O}\right) = 6.04 \text{ g H}$$

34. $(1.00 \text{ L})\left(\frac{1000 \text{ mL}}{1 \text{ L}}\right)\left(\frac{1.55 \text{ g}}{1.00 \text{ mL}}\right)\left(\frac{0.650 \text{ g } H_2SO_4}{1.00 \text{ g}}\right)\left(\frac{1 \text{ mol}}{98.08 \text{ g}}\right) = 10.3 \text{ mol } H_2SO_4$

35. $(100. \text{ mL})\left(\frac{1.42 \text{ g}}{\text{mL}}\right)\left(\frac{0.720 \text{ g } HNO_3}{1.000 \text{ g}}\right)\left(\frac{1 \text{ mol}}{63.02 \text{ g } HNO_3}\right) = 1.62 \text{ mol } HNO_3$

36. (a) Determine the molar mass of each compound.

CO_2, 44.01 g; O_2, 32.00 g; H_2O, 18.02 g; CH_3OH, 32.04 g. The 1.00 gram sample with the lowest molar mass will contain the most moleclues. Thus, H_2O will contain the most molecuels.

(b) $(1.00 \text{ g } H_2O)\left(\frac{1 \text{ mol}}{18.02 \text{ g}}\right)\left(\frac{(3) \ 6.022 \times 10^{23} \text{ atoms}}{\text{mol}}\right) = 1.00 \times 10^{23} \text{ atoms}$

$(1.00 \text{ g } CH_3OH)\left(\frac{1 \text{ mol}}{32.04 \text{ g}}\right)\left(\frac{(6) \ 6.022 \times 10^{23} \text{ atoms}}{\text{mole}}\right) = 1.13 \times 10^{23} \text{ atoms}$

The 1.00 g sample of CH_3OH contains the most atoms

37. 1 mol Fe_2S_3 = 207.9 g Fe_2S_3 = 6.022×10^{23} formula units

$(6.022 \times 10^{23} \text{ atoms})\left(\frac{207.9 \text{ g } Fe_2S_3}{6.022 \times 10^{23} \text{ formula units}}\right)\left(\frac{1 \text{ formula unit}}{5 \text{ atoms}}\right) = 41.58 \text{ g } Fe_2S_3$

38. Percent composition

(a) NaBr   Na 22.99 $\left(\frac{22.99}{102.9}\right)(100)$ = 22.34% Na
             Br 79.90 $\left(\frac{79.90}{102.9}\right)(100)$ = 77.65% Br
                 102.9

(b) $KHCO_3$ K 39.10 $\left(\frac{39.10}{100.1}\right)(100)$ = 39.06% K
                 H 1.008 $\left(\frac{1.008}{100.1}\right)(100)$ = 1.007% H
             3   O 48.00
                 C 12.01 $\left(\frac{12.01}{100.1}\right)(100)$ = 12.00% C
                     100.1 $\left(\frac{48.00}{100.1}\right)(100)$ = 47.95% O

(c) $FeCl_3$ Fe 55.85 $\left(\frac{55.85}{162.3}\right)(100)$ = 34.41% Fe
             3 Cl 106.4 $\left(\frac{106.4}{162.3}\right)(100)$ = 65.56% Cl
                     162.3

(d) $SiCl_4$

	Si	28.09	$\left(\frac{28.09}{169.9}\right)(100)$ = 16.53% Si
4	Cl	141.8	$\left(\frac{141.8}{169.9}\right)(100)$ = 83.46% Cl
		169.9	

(e) $Al_2(SO_4)_3$

2	Al	53.96	$\left(\frac{53.96}{342.1}\right)(100)$ = 15.77% Al
3	S	96.18	$\left(\frac{96.18}{342.1}\right)(100)$ = 28.11% S
12	O	192.0	$\left(\frac{192.0}{342.1}\right)(100)$ = 56.12% O
		342.1	

(f) $AgNO_3$

	Ag	107.9	$\left(\frac{107.9}{169.9}\right)(100)$ = 63.51% Ag
	N	14.01	$\left(\frac{14.01}{169.9}\right)(100)$ = 8.246% N
3	O	48.00	$\left(\frac{48.00}{169.9}\right)(100)$ = 28.25% O
		169.9	

39. Percent composition

(a) $ZnCl_2$

	Zn	65.38	$\left(\frac{65.38}{136.3}\right)(100)$ = 47.97% Zn
2	Cl	70.90	$\left(\frac{70.90}{136.3}\right)(100)$ = 52.02% Cl
		136.3	

(b) $NH_4C_2H_3O_2$

	N	14.01	$\left(\frac{14.01}{77.09}\right)(100)$ = 18.17% N
7	H	7.056	$\left(\frac{7.056}{77.09}\right)(100)$ = 9.153% H
2	C	24.02	$\left(\frac{24.02}{77.09}\right)(100)$ = 31.16% C
2	O	32.00	$\left(\frac{32.00}{77.09}\right)(100)$ = 41.51% O
		77.09	

(c) MgP_2O_7

	Mg	24.31	$\left(\frac{24.31}{198.3}\right)(100)$ = 12.26% Mg
2	P	61.94	$\left(\frac{61.94}{198.3}\right)(100)$ = 31.24% P
7	O	112.0	$\left(\frac{112.0}{198.3}\right)(100)$ = 56.48% O
		198.3	

(d) $(NH_4)_2SO_4$

2	N	28.02	$\left(\frac{28.02}{132.1}\right)(100)$ = 21.21% N
8	H	8.064	$\left(\frac{8.064}{132.1}\right)(100)$ = 6.104% H
	S	32.06	$\left(\frac{32.06}{132.1}\right)(100)$ = 24.27% S
4	O	64.00	$\left(\frac{64.00}{132.1}\right)(100)$ = 48.45% O
		132.1	

(e) $Fe(NO_3)_3$ Fe   55.85 $\left(\frac{55.85}{241.9}\right)(100)$ =   23.09% Fe

     3   N   42.03

     9   O   144.0 $\left(\frac{42.03}{241.9}\right)(100)$ =   17.37% N

          $\overline{241.9}$

               $\left(\frac{144.0}{241.9}\right)(100)$ =   59.53%   O

(f) ICl_3 I   126.9 $\left(\frac{126.9}{233.3}\right)(100)$ =   54.39% I

     3   Cl   106.4 $\left(\frac{106.4}{233.3}\right)(100)$ =   45.61% Cl

         $\overline{233.3}$

40. Percent of iron

(a) FeO   Fe   55.85 $\left(\frac{55.85}{71.85}\right)(100)$ =   77.73% Fe

        O   16.00

         $\overline{71.85}$

(b) Fe_2O_3 2   Fe   111.7 $\left(\frac{111.7}{159.7}\right)(100)$ =   69.94% Fe

     3   O   48.00

         $\overline{159.7}$

(c) Fe_3O_4 3   Fe   167.4 $\left(\frac{167.6}{231.6}\right)(100)$ =   72.37% Fe

     4   O   64.00

         $\overline{231.6}$

(d) $K_4Fe(CN)_6$ Fe   55.85 $\left(\frac{55.85}{368.4}\right)(100)$ =   15.16% Fe

     4   K   156.4

     6   C   72.06

     6   N   84.06

         $\overline{368.4}$

41. Percent chlorine

(a) KCl   K   39.10 $\left(\frac{35.45}{74.55}\right)(100)$ =   47.62% Cl

       Cl   35.45

        $\overline{74.55}$

(b) $BaCl_2$ Ba   137.3 $\left(\frac{70.90}{208.2}\right)(100)$ =   34.05% Cl

     2   Cl   70.90

        $\overline{208.2}$

(c) $SiCl_4$ Si   28.09 $\left(\frac{141.8}{169.9}\right)(100)$ =   83.46% Cl

     4   Cl   141.8

        $\overline{169.9}$

(d) LiCl   Li   6.941 $\left(\frac{35.45}{42.39}\right)(100)$ =   83.63% Cl

       Cl $\underline{35.45}$

         42.39

Highest % Cl is LiCl: lowest % Cl is $BaCl_2$

42. Percent composition of oxide

14.20 g oxide
−6.20 g P
8.00 g oxygen

$\left(\frac{6.20}{14.20}\right)(100) = 43.7\%$ P

$\left(\frac{8.00}{14.20}\right)(100) = 56.3\%$ O

43. Percent composition of ethylene chloride

6.00 g C
1.00 g H
17.75 g Cl
24.75 g total

$\left(\frac{6.00}{24.75}\right)(100) = 24.2\%$ C

$\left(\frac{1.00}{24.75}\right)(100) = 4.00\%$ H

$\left(\frac{17.75}{24.75}\right)(100) = 71.72\%$ Cl

44. From the formula, 2 Li (13.88 g) combine with 1 S (32.06 g).

$\left(\frac{13.88 \text{ g Li}}{32.06 \text{ g S}}\right)(20.0 \text{ g S}) = 8.66 \text{ g Li}$

45. (a) $HgCO_3$

	Hg	200.6
	C	12.01
3	O	48.00
		260.6

$\left(\frac{200.6}{260.6}\right)(100) = 76.98\%$ Hg

(b) $Ca(ClO_3)_2$

6	O	96.00
2	Cl	70.90
	Ca	40.08
		207.0

$\left(\frac{96.00}{207.0}\right)(100) = 46.38\%$ O

(c) $C_{10}H_{14}N_2$

2	N	28.02
10	C	120.1
14	H	14.11
		162.2

$\left(\frac{28.02}{162.2}\right)(100) = 17.27\%$ N

(d) $C_{55}H_{72}MgN_4O_5$

	Mg	24.31
55	C	660.55
72	H	72.58
4	N	56.04
5	O	80.00
		892.5

$\left(\frac{24.31}{892.5}\right)(100) = 2.721\%$ Mg

46. (a) H_2O (c) equal   (e) $KHSO_4$

(b) N_2O_3 (d) $KClO_3$ (f) Na_2CrO_4

47. Empirical formulas from percent composition.

    (a)   Step 1.   Express each element as grams/100 g material

$$63.6\% \text{ N} = 63.6 \text{ g N/100 g material}$$
$$36.4\% \text{ O} = 36.4 \text{ g O/100 g material}$$

           Step 2.   Calculate the relative moles of each element.

$$(63.6 \text{ g N}) \left(\frac{1 \text{ mol}}{14.01 \text{ g}}\right) = 4.54 \text{ mol N}$$
$$(36.4 \text{ g O}) \left(\frac{1 \text{ mol}}{16.00 \text{ g}}\right) = 2.28 \text{ mol O}$$

           Step 3.   Change these moles to whole numbers by dividing each by the smaller number.

$$\frac{4.54 \text{ mol N}}{2.28} = 1.99 \text{ mol N}$$
$$\frac{2.28 \text{ mol O}}{2.28} = 1.00 \text{ mol O}$$

The simplest ratio of N:O is 2:1. The empirical formula, therfore, is N_2O

    (b)   46.7% N, 53.3% O

$$(46.7 \text{ g N})\left(\frac{1 \text{ mol}}{14.01 \text{ g}}\right) = 3.34 \text{ mol N} \qquad \frac{3.34}{3.33} = 1.00 \text{ mol N}$$
$$(53.3 \text{ g O})\left(\frac{1 \text{ mol}}{16.00 \text{ g}}\right) = 3.33 \text{ mol O} \qquad \frac{3.33}{3.33} = 1.00 \text{ mol O}$$

The empirical formula is NO.

    (c)   25.9% N, 71.4% O

$$(25.9 \text{ g N})\left(\frac{1 \text{ mol}}{14.01 \text{ g}}\right) = 1.85 \text{ mol N} \qquad \frac{1.85}{1.85} = 1.00 \text{ mol N}$$
$$(74.1 \text{ g O})\left(\frac{1 \text{ mol}}{16.00 \text{ g}}\right) = 4.63 \text{ mol O} \qquad \frac{4.63}{1.85} = 2.5 \text{ mol O}$$

Since these values are not whole numbers, mutiply each by 2 to change them to whole numbers.

$$(1.00 \text{ mol N})(2) = 2.00 \text{ mol N}; \quad (2.5 \text{ mol O})(2) = 5.00 \text{ mol O}$$

The empirical formula is N_2O_5.

    (d)   43.4% Na, 11.3% C, 45.3% O

$$(43.4 \text{ g Na})\left(\frac{1 \text{ mol}}{22.99 \text{ g}}\right) = 1.89 \text{ mol Na} \qquad \frac{1.89}{0.942} = 2.01 \text{ mol Na}$$
$$(11.3 \text{ g C})\left(\frac{1 \text{ mol}}{12.01 \text{ g}}\right) = 0.942 \text{ mol C} \qquad \frac{0.942}{0.942} = 1.00 \text{ mol C}$$
$$(45.3 \text{ g O})\left(\frac{1 \text{ mol}}{16.00 \text{ g}}\right) = 2.83 \text{ mol O} \qquad \frac{2.83}{0.942} = 3.00 \text{ mol O}$$

The empirical formula is Na_2CO_3.

(e)   18.8 % Na, 29.0% Cl, 52.3% O

$$(18.8 \text{ g Na})\left(\frac{1 \text{ mol}}{22.99 \text{ g}}\right) = 0.818 \text{ mol Na} \qquad \frac{0.818}{0.818} = 1.00 \text{ mol Na}$$

$$(29.0 \text{ g Cl})\left(\frac{1 \text{ mol}}{35.45 \text{ g}}\right) = 0.818 \text{ mol Cl} \qquad \frac{0.818}{0.818} = 1.00 \text{ mol Cl}$$

$$(52.3 \text{ g O})\left(\frac{1 \text{ mol}}{16.00 \text{ g}}\right) = 3.27 \text{ mol O} \qquad \frac{3.27}{0.818} = 4.00 \text{ mol O}$$

The empirical formula is $NaClO_4$.

(f)   72.02% Mn, 27.98% O

$$(72.02 \text{ g Mn})\left(\frac{1 \text{ mol}}{54.94 \text{ g}}\right) = 1.31 \text{ mol Mn} \qquad \frac{1.31}{1.31} = 1.00 \text{ mol Mn}$$

$$(27.98 \text{ g O})\left(\frac{1 \text{ mol}}{16.00 \text{ g}}\right) = 1.75 \text{ mol O} \qquad \frac{1.75}{1.31} = 1.34 \text{ mol O}$$

Multiply both values by 3 to give whole numbers.

(1.00 mol Mn)(3) = 3.00 mol Mn; (1.34 mol O)(3) = 4.02 mol O

The empirical formula is Mn_3O_4.

48.   Empirical formulas from percent composition

(a)   64.1% Cu, 35.9% Cl

$$(64.1 \text{ g Cu})\left(\frac{1 \text{ mol}}{63.55 \text{ g}}\right) = 1.01 \text{ mol Cu} \qquad \frac{1.01}{1.01} = 1.00 \text{ mol Cu}$$

$$(35.9 \text{ g Cl})\left(\frac{1 \text{ mol}}{35.45 \text{ g}}\right) = 1.01 \text{ mol Cl} \qquad \frac{1.01}{1.01} = 1.00 \text{ mol Cl}$$

The empirical formula is CuCl.

(b)   47.2% Cu, 52.8% Cl

$$(47.2 \text{ g Cu})\left(\frac{1 \text{ mol}}{63.55 \text{ g}}\right) = 0.743 \text{ mol Cu} \qquad \frac{0.743}{0.743} = 1.00 \text{ mol Cu}$$

$$(52.8 \text{ g Cl})\left(\frac{1 \text{ mol}}{35.45 \text{ g}}\right) = 1.49 \text{ mol Cl} \qquad \frac{1.49}{0.743} = 2.00 \text{ mol Cl}$$

The empirical formula is $CuCl_2$.

(c)   51.9% Cr, 48.1% S

$$(51.9 \text{ g Cr})\left(\frac{1 \text{ mol}}{52.00 \text{ g}}\right) = 0.998 \text{ mol Cr} \qquad \frac{0.998}{0.998} = 1.00 \text{ mol Cr}$$

$$(48.1 \text{ g S})\left(\frac{1 \text{ mol}}{32.06 \text{ g}}\right) = 1.50 \text{ mol S} \qquad \frac{1.50}{0.998} = 1.50 \text{ mol S}$$

Multiply both values by 2 to give whole numbers.

(1.00 mol Cr)(2) = 2.00 mol Cr; (1.50 mol S)(2) = 3.00 mol S

The empirical formula is Cr_2S_3.

(d)   55.3% K, 14.6% P, 30.1% O

$$(55.3 \text{ g K})\left(\frac{1 \text{ mol}}{39.10 \text{ g}}\right) = 1.41 \text{ mol K} \qquad \frac{1.41}{0.471} = 2.99 \text{ mol K}$$

$$(14.6 \text{ g P})\left(\frac{1 \text{ mol}}{30.97 \text{ g}}\right) = 0.471 \text{ mol P} \qquad \frac{0.471}{0.471} = 1.00 \text{ mol P}$$

$$(30.1 \text{ g O})\left(\frac{1 \text{ mol}}{16.00 \text{ g}}\right) = 1.88 \text{ mol O} \qquad \frac{1.88}{0.471} = 3.99 \text{ mol O}$$

The empirical formula is K_3PO_4.

(e)   38.9% Ba, 29.4% Cr, 31.7% O

$$(38.9 \text{ g Ba})\left(\frac{1 \text{ mol}}{137.3 \text{ g}}\right) = 0.283 \text{ mol Ba} \qquad \frac{0.283}{0.283} = 1.00 \text{ mol Ba}$$

$$(29.4 \text{ g Cr})\left(\frac{1 \text{ mol}}{52.00 \text{ g}}\right) = 0.565 \text{ mol Cr} \qquad \frac{0.565}{0.283} = 1.99 \text{ mol Cr}$$

$$(31.7 \text{ g O})\left(\frac{1 \text{ mol}}{16.00 \text{ g}}\right) = 1.98 \text{ mol O} \qquad \frac{1.98}{0.283} = 7.00 \text{ mol O}$$

The empirical formula is $BaCr_2O_7$.

(f)   3.99% P, 82.3% Br, 13.7% Cl

$$(3.99 \text{ g P})\left(\frac{1 \text{ mol}}{30.97 \text{ g}}\right) = 0.129 \text{ mol P} \qquad \frac{0.129}{0.129} = 1.00 \text{ mol P}$$

$$(82.3 \text{ g Br})\left(\frac{1 \text{ mol}}{79.90 \text{ g}}\right) = 1.03 \text{ mol Br} \qquad \frac{1.03}{0.129} = 7.98 \text{ mol Br}$$

$$(13.7 \text{ g Cl})\left(\frac{1 \text{ mol}}{35.45 \text{ g}}\right) = 0.386 \text{ mol Cl} \qquad \frac{0.386}{0.129} = 2.99 \text{ mol Cl}$$

The empirical formula is PBr_8Cl_3.

49. Empirical formula

$$(3.996 \text{ g Sn})\left(\frac{1 \text{ mol}}{118.7 \text{ g}}\right) = 0.0337 \text{ mol Sn} \qquad \frac{0.0337}{0.0337} = 1.00 \text{ mol Sn}$$

$$(1.077 \text{ g O})\left(\frac{1 \text{ mol}}{16.00 \text{ g}}\right) = 0.0673 \text{ mol O} \qquad \frac{0.0673}{0.0337} = 2.00 \text{ mol O}$$

The empirical formula is SnO_2.

50. Empirical formula

5.454 g product − 3.054 g V = 2.400 g O

$$(3.054 \text{ g V})\left(\frac{1 \text{ mol}}{50.94 \text{ g}}\right) = 0.0600 \text{ mol V} \qquad \frac{0.0600}{0.0600} = 1.00 \text{ mol V}$$

$$(2.400 \text{ g O})\left(\frac{1 \text{ mol}}{16.00 \text{ g}}\right) = 0.1500 \text{ mol O} \qquad \frac{0.1500}{0.0600} = 2.50 \text{ mol O}$$

Multiplying both by 2 gives the empirical formula V_2O_5.

51. According to the formula, 1 mol (65.4 g) Zn combines with 1 mol (32.1 g) S.

$$(19.5 \text{ g Zn})\left(\frac{32.06 \text{ g S}}{65.38 \text{ g Zn}}\right) = 9.56 \text{ g S}$$

19.5 g Zn require 9.57 g S for complete reaction. Therefore, there is not sufficient S present (9.40 g) to react with the Zn.

52. 65.45% C, 5.45% H, 29.09% O; molar mass = 110.1

$$(65.45 \text{ g C})\left(\frac{1 \text{ mol}}{12.01 \text{ g}}\right) = 5.450 \text{ mol C} \qquad \frac{5.450}{1.818} = 2.998 \text{ mol C}$$

$$(5.45 \text{ g H})\left(\frac{1 \text{ mol}}{1.008 \text{ g}}\right) = 5.41 \text{ mol H} \qquad \frac{5.41}{1.818} = 2.98 \text{ mol H}$$

$$(29.09 \text{ g O})\left(\frac{1 \text{ mol}}{16.00 \text{ g}}\right) = 1.818 \text{ mol O} \qquad \frac{1.818}{1.818} = 1.000 \text{ mol O}$$

The empirical formula is C_3H_3O making the empirical mass 55.05.

$$\frac{\text{molar mass}}{\text{empirical mass}} = \frac{110.1}{55.05} = 2$$

The molecular formula is twice that of the empirical formula.
Molecular formula = $(C_3H_3O)_2 = C_6H_6O_2$

53. 40.0% C, 6.7% H, 53.3% O; molar mass = 180.1

$$(40.0 \text{ g C})\left(\frac{1 \text{ mol}}{12.01 \text{ g}}\right) = 3.33 \text{ mol C} \qquad \frac{3.33}{3.33} = 1.00 \text{ mol C}$$

$$(6.7 \text{ g H})\left(\frac{1 \text{ mol}}{1.008 \text{ g}}\right) = 6.6 \text{ mol H} \qquad \frac{6.6}{3.33} = 1.99 \text{ mol H}$$

$$(53.3 \text{ g O})\left(\frac{1 \text{ mol}}{16.00 \text{ g}}\right) = 3.33 \text{ mol O} \qquad \frac{3.33}{3.33} = 1.00 \text{ mol O}$$

The empirical formula is CH_2O making the empirical mass 30.03.

$$\frac{\text{molar mass}}{\text{empirical mass}} = \frac{180.1}{30.03} = 6$$

The molecular formula is six times that of the empirical formula.

Molecular formula = $(CH_2O)_6 = C_6H_{12}O_6$

54. 60.0% C, 4.48% H, 35.5% O; molar mass of aspirin = 180.2

$$(60.0 \text{ g C})\left(\frac{1 \text{ mol}}{12.01 \text{ g}}\right) = 5.00 \text{ mol C} \qquad \frac{5.00}{2.22} = 2.25 \text{ mol C}$$

$$(4.48 \text{ g H})\left(\frac{1 \text{ mol}}{1.008 \text{ g}}\right) = 4.44 \text{ mol H} \qquad \frac{4.44}{2.22} = 2.00 \text{ mol H}$$

$$(35.5 \text{ g O})\left(\frac{1 \text{ mol}}{16.00 \text{ g}}\right) = 2.22 \text{ mol O} \qquad \frac{2.22}{2.22} = 1.00 \text{ mol O}$$

Multiplying each by 4 gives the empirical formula $C_9H_8O_4$. The empirical mass is 180.2 Since the empirical mass equals the molar mass, the molecular formula is the same as the empirical formula, $C_9H_8O_4$.

55. Calculate the percent of oxygen in $Al_2(SO_4)_3$.

2	Al	53.96
3	S	96.18
12	O	192.0
		342.1

$\left(\frac{192.0}{342.1}\right)(100) = 56.12\%$ O

Now take 56.12% of 8.50 g.

$(8.50 \text{ g O})(0.5612) = 4.77 \text{ g O}$

56. Empirical formula of gallium arsenide; 48.2% Ga, 51.8% As

$(48.2 \text{ g Ga})\left(\frac{1 \text{ mol}}{69.72 \text{ g}}\right) = 0.691 \text{ mol Ga} \qquad \frac{0.691}{0.691} = 1.00 \text{ mol Ga}$

$(51.8 \text{ g As})\left(\frac{1 \text{ mol}}{74.92 \text{ g}}\right) = 0.691 \text{ mol As} \qquad \frac{0.691}{0.691} = 1.00 \text{ mol As}$

The empirical formula is GaAs.

57. (a) 7.79% C, 92.21% Cl

$(7.79 \text{ g C})\left(\frac{1 \text{ mol}}{12.01 \text{ g}}\right) = 0.649 \text{ mol C} \qquad \frac{0.649}{0.649} = 1.00 \text{ mol C}$

$(92.21 \text{ g Cl})\left(\frac{1 \text{ mol}}{35.45 \text{ g}}\right) = 2.601 \text{ mol Cl} \qquad \frac{2.601}{0.649} = 4.01 \text{ mol Cl}$

The empirical formula is CCl_4. The empirical mass is 153.8 which equals the molar mass, therefore the molecular formula is CCl_4.

(b) 10.13% C, 89.87% Cl

$(10.13 \text{ g C})\left(\frac{1 \text{ mol}}{12.01 \text{ g}}\right) = 0.8435 \text{ mol C} \qquad \frac{0.8435}{0.8435} = 1.00 \text{ mol C}$

$(89.87 \text{ g Cl})\left(\frac{1 \text{ mol}}{35.45 \text{ g}}\right) = 2.535 \text{ mol Cl} \qquad \frac{2.535}{0.8435} = 3.00 \text{ mol Cl}$

The empirical formula is CCl_3. The empirical mass is 118.4.

$\frac{\text{molar mass}}{\text{empirical mass}} = \frac{236.7}{118.4} = 2$

The molecular formula is twice that of the empirical formula.

Molecular formula $= C_2Cl_6$.

(c) 25.26% C, 74.74% Cl

$(25.26 \text{ g C})\left(\frac{1 \text{ mol}}{12.01 \text{ g}}\right) = 2.103 \text{ mol C} \qquad \frac{2.103}{2.103} = 1.00 \text{ mol C}$

$(74.74 \text{ g Cl})\left(\frac{1 \text{ mol}}{35.45 \text{ g}}\right) = 2.108 \text{ mol Cl} \qquad \frac{2.108}{2.103} = 1.00 \text{ mol Cl}$

The empirical formula is CCl. The empirical mass is 47.46.

$\frac{\text{molar mass}}{\text{empirical mass}} = \frac{284.8}{47.46} = 6$

The molecular formula is six times that of the empirical formula.

Molecular formula $= C_6Cl_6$.

(d)   11.25% C, 88.75% Cl

$(11.25 \text{ g C})\left(\frac{1 \text{ mol}}{12.01 \text{ g}}\right)$ =   0.9367 mol C $\frac{0.9367}{0.9367} = 1.00$ mol C

$(88.75 \text{ g Cl})\left(\frac{1 \text{ mol}}{35.45 \text{ g}}\right)$ =   2.504 mol Cl $\frac{2.504}{0.9367} = 2.67$ mol Cl

Multiplying each by 3 gives the empirical formula C_3Cl_8.   The empirical mass is 319.6 Since the molar mass is also 319.6 the molecular formula is C_3Cl_8.

CHEMICAL EQUATIONS

1. (a) $2\,H_2 + O_2 \longrightarrow 2\,H_2O$

     (b) $H_2 + Br_2 \longrightarrow 2\,HBr$

     (c) $3\,C + Fe_2O_3 \longrightarrow 2\,Fe + 3\,CO$

     (d) $2\,H_2O_2 \longrightarrow 2\,H_2O + O_2$

     (e) $Ba(ClO_3)_2 \xrightarrow{\Delta} BaCl_2 + 3\,O_2$

     (f) $H_2SO_4 + 2\,NaOH \longrightarrow 2\,H_2O + Na_2SO_4$

     (g) $2\,NH_4I + Cl_2 \longrightarrow 2\,NH_4Cl + I_2$

     (h) $CrCl_3 + 3\,AgNO_3 \longrightarrow Cr(NO_3)_3 + 3\,AgCl$

     (i) $Al_2(CO_3)_3 \xrightarrow{\Delta} Al_2O_3 + 3\,CO_2$

     (j) $4\,Al + 3\,C \xrightarrow{\Delta} Al_4C_3$

2. (a)   combination   (f)   double displacement

     (b)   combination   (g)   single displacement

     (c)   single displacement   (h)   double displacement

     (d)   decomposition   (i)   decomposition

     (e)   decomposition   (j)   combination

3. (a) $2\,SO_2 + O_2 \longrightarrow 2\,SO_3$

     (b) $4\,Al + 3\,MnO_2 \xrightarrow{\Delta} 3\,Mn + 2\,Al_2O_3$

     (c) $2\,Na + 2\,H_2O \longrightarrow 2\,NaOH + H_2$

     (d) $2\,AgNO_3 + Ni \longrightarrow Ni(NO_3)_2 + 2\,Ag$

     (e) $Bi_2S_3 + 6\,HCl \longrightarrow 2\,BiCl_3 + 3\,H_2S$

     (f) $2\,PbO_2 \xrightarrow{\Delta} 2\,PbO + O_2$

     (g) $2\,LiAlH_4 \xrightarrow{\Delta} 2\,LiH + 2\,Al + 3\,H_2$

     (h) $2\,KI + Br_2 \longrightarrow 2\,KBr + I_2$

     (i) $2\,K_3PO_4 + 3\,BaCl_2 \longrightarrow 6\,KCl + Ba_3(PO_4)_2$

4. (a) $2\,MnO_2 + CO \longrightarrow Mn_2O_3 + CO_2$

     (b) $Mg_3N_2 + 6\,H_2O \longrightarrow 3\,Mg(OH)_2 + 2\,NH_3$

     (c) $4\,C_3H_5(NO_3)_3 \longrightarrow 12\,CO_2 + 10\,H_2O + 6\,N_2 + O_2$

     (d) $4\,FeS + 7\,O_2 \longrightarrow 2\,Fe_2O_3 + 4\,SO_2$

(e) $2 Cu(NO_3)_2 \longrightarrow 2 CuO + 4 NO_2 + O_2$

(f) $3 NO_2 + H_2O \longrightarrow 2 HNO_3 + NO$

(g) $2 Al + 3 H_2SO_4 \longrightarrow Al_2(SO_4)_3 + 3 H_2$

(h) $4 HCN + 5 O_2 \longrightarrow 2 N_2 + 4 CO_2 + 2 H_2O$

(i) $2 B_5H_9 + 12 O_2 \longrightarrow 5 B_2O_3 + 9 H_2O$

(j) $4 NH_3 + 5 O_2 \longrightarrow 4 NO + 6 H_2O$

5. (a) $2 Cu + S \longrightarrow Cu_2S$

   (b) $2 H_3PO_4 + 3 Ca(OH)_2 \xrightarrow{\Delta} Ca_3(PO_4)_2 + 6 H_2O$

   (c) $2 Ag_2O \xrightarrow{\Delta} 4 Ag + O_2$

   (d) $FeCl_3 + 3 NaOH \longrightarrow Fe(OH)_3 + 3 NaCl$

   (e) $Ni_3(PO_4)_2 + 3 H_2SO_4 \longrightarrow 3 NiSO_4 + 2 H_3PO_4$

   (f) $ZnCO_3 + 2 HCl \longrightarrow ZnCl_2 + H_2O + CO_2$

   (g) $3 AgNO_3 + AlCl_3 \longrightarrow 3 AgCl + Al(NO_3)_3$

6. (a) $2 H_2O \longrightarrow 2 H_2 + O_2$

   (b) $HC_2H_3O_2 + KOH \longrightarrow KC_2H_3O_2 + H_2O$

   (c) $2 P + 3 I_2 \longrightarrow 2 PI_3$

   (d) $2 Al + 3 CuSO_4 \longrightarrow 3 Cu + Al_2(SO_4)_3$

   (e) $(NH_4)_2SO_4 + BaCl_2 \longrightarrow 2 NH_4Cl + BaSO_4$

   (f) $SF_4 + 2 H_2O \longrightarrow SO_2 + 4 HF$

   (g) $Cr_2(CO_3)_3 \xrightarrow{\Delta} Cr_2O_3 + 3 CO_2$

7. (a) $2 K + O_2 \longrightarrow 2 K_2O$

   (b) $2 Al + 3 Cl_2 \longrightarrow 2 AlCl_3$

   (c) $CO_2 + H_2O \longrightarrow H_2CO_3$

   (d) $CaO + H_2O \longrightarrow Ca(OH)_2$

8. (a) $2 HgO \xrightarrow{\Delta} 2 Hg + O_2\uparrow$

   (b) $2 NaClO_3 \xrightarrow{\Delta} 2 NaCl + 3 O_2\uparrow$

   (c) $MgCO_3 \xrightarrow{\Delta} MgO + CO_2\uparrow$

   (d) $2 PbO_2 \xrightarrow{\Delta} 2 PbO + O_2\uparrow$

9. (a) $Zn + H_2SO_4 \longrightarrow H_2\uparrow + ZnSO_4$

   (b) $2 AlI_3 + 3 Cl_2 \longrightarrow 2 AlCl_3 + 3 I_2$

   (c) $Mg + 2 AgNO_3 \longrightarrow Mg(NO_3)_2 + 2 Ag$

   (d) $2 Al + 3 CoSO_4 \longrightarrow Al_2(SO_4)_3 + 3 Co$

10. (a) $Ag + H_2SO_4(aq) \longrightarrow$ no reaction

    (b) $Cl_2 + 2 NaBr(aq) \longrightarrow Br_2 + 2 NaCl(aq)$

    (c) $Mg + ZnCl_2(aq) \longrightarrow Zn + MgCl_2(aq)$

    (d) $Pb + 2 AgNO_3(aq) \longrightarrow 2 Ag + Pb(NO_3)_2(aq)$

    (e) $Cu + FeCl_3(aq) \longrightarrow$ no reaction

    (f) $H_2 + Al_2O_3(s) \overset{\Delta}{\longrightarrow}$ no reaction

    (g) $2 Al + 6 HBr(aq) \longrightarrow 3 H_2(g) + 2 AlBr_3(aq)$

    (h) $I_2 + HCl(aq) \longrightarrow$ no reaction

11. (a) $ZnCl_2 + 2 KOH \longrightarrow Zn(OH)_2 + 2 KCl$

    (b) $CuSO_4 + H_2S \longrightarrow H_2SO_4 + CuS$

    (c) $3 Ca(OH)_2 + 2 H_3PO_4 \longrightarrow 6 H_2O + Ca_3(PO_4)_2$

    (d) $2 (NH_4)_3PO_4 + 3 Ni(NO_3)_2 \longrightarrow 6 NH_4NO_3 + Ni_3(PO_4)_2$

    (e) $Ba(OH)_2 + 2 HNO_3 \longrightarrow H_2O + Ba(NO_3)_2$

    (f) $(NH_4)_2S + 2 HCl \longrightarrow H_2S + 2 NH_4Cl$

12. (a) $AgNO_3(aq) + KCl(aq) \longrightarrow AgCl\downarrow + KNO_3(aq)$

    (b) $Ba(NO_3)_2(aq) + MgSO_4(aq) \longrightarrow Mg(NO_3)_2(aq) + BaSO_4\downarrow$

    (c) $H_2SO_4(aq) + Mg(OH)_2(aq) \longrightarrow 2 H_2O + MgSO_4(aq)$

    (d) $MgO(s) + H_2SO_4(aq) \longrightarrow H_2O + MgSO_4(aq)$

    (e) $Na_2CO_3(aq) + NH_4Cl(aq) \longrightarrow$ no reaction

13. (a) $H_2 + I_2 \longrightarrow 2 HI$

    (b) $CaCO_3 \overset{\Delta}{\longrightarrow} CaO + CO_2\uparrow$

    (c) $Mg + H_2SO_4 \longrightarrow H_2\uparrow + MgSO_4$

    (d) $FeCl_2 + NaOH \longrightarrow Fe(OH)_2 + 2 NaCl$

    (e) $SO_2 + H_2O \longrightarrow H_2SO_3$

    (f) $SO_3 + H_2O \longrightarrow H_2SO_4$

    (g) $Ca + H_2O \longrightarrow Ca(OH)_2 + H_2\uparrow$

    (h) $2 Bi(NO_3)_3 + 3 H_2S \longrightarrow Bi_2S_3 + 6 HNO_3$

14. (a) $2 Ba + O_2 \longrightarrow 2 BaO$

    (b) $2 NaHCO_3 \overset{\Delta}{\longrightarrow} Na_2CO_3 + H_2O + CO_2\uparrow$

    (c) $Ni + CuSO_4 \longrightarrow NiSO_4 + Cu$

    (d) $MgO + 2 HCl \longrightarrow MgCl_2 + H_2O$

    (e) $H_3PO_4 + 3 KOH \longrightarrow K_3PO_4 + 3 H_2O$

(f) $C + O_2 \xrightarrow{\Delta} CO_2\uparrow$

(g) $2\ Al(ClO_3)_3 \xrightarrow{\Delta} 9\ O_2\uparrow + 2\ AlCl_3$

(h) $CuBr_2 + Cl_2 \longrightarrow CuCl_2 + Br_2$

(i) $2\ SbCl_3 + 3\ (NH_4)_2S \longrightarrow Sb_2S_3 + 6\ NH_4Cl$

(j) $2\ NaNO_3 \xrightarrow{\Delta} 2\ NaNO_2 + O_2\uparrow$

15.   The purpose of balancing chemical equations is to conform to the Law of Conservation of Mass. Ratios of reactants and products can then be easily determined.

16.   The coefficients in a balanced chemical equation represent the number of moles (or molecules or formula units) of each of the chemical species in the reaction.

17.   (a)   One mole of $MgBr_2$ reacts with two moles of $AgNO_3$ to yield one mole of $Mg(NO_3)_2$ and two moles of AgBr.

      (b)   One mole of N_2 reacts with three moles of H_2 to produce two moles of NH_3.

      (c)   Two moles of C_3H_7OH react with nine moles O_2 to form six moles of CO_2 and eight moles of H_2O.

18.   (a)   Two moles of Na react with one mole of Cl_2 to produce two moles of NaCl and release 822 kJ of energy.   The reaction is exothermic.

      (b)   One mole of PCl_5 absorbs 92.9 kJ of energy to produce one mole of PCl_3 and one mole of Cl_2.   The reaction is endothermic.

19.   (a) $CaO + H_2O \longrightarrow Ca(OH)_2 + 65.3\ kJ$

      (b) $2\ Al_2O_3 + 3260\ kJ \longrightarrow 4\ Al + 3\ O_2$

20.   (a) $2\ C_2H_6 + 7\ O_2 \longrightarrow 4\ CO_2 + 6\ H_2O$

      (b) $2\ C_6H_6 + 15\ O_2 \longrightarrow 12\ CO_2 + 6\ H_2O$

      (c) $C_7H_{16} + 11\ O_2 \longrightarrow 7\ CO_2 + 8\ H_2O$

21.   The correct statements are a, d, e, f, h, i, j, l, n, o, p, q.

      (b)   A balanced chemical equation is one that has the same number of atoms of each element on each side of the equation.

      (c)   In a chemical equation, the symbol $\xrightarrow{\Delta}$ indicates that heat is required during the reaction.

      (g)   The equation $H_2O \longrightarrow H_2 + O_2$ can be balanced by placing a 2 in front of the H_2O and a 2 in front of the H_2.

      (k)   The products are the substances produced by the chemical reaction.

      (m)   When a precipitate is formed in a chemical reaction, it can be indicated in the equation with the symbol (S) or ↓ immediately following the formula of the substance precipitated.

22. pigment   function

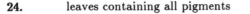

    chlorophyll   photosynthesis (energy absorption)
    carotenoids   absorb high energy oxygen and releases it appropriately
    anthocyanins   diversity in leaf color

23. CO_2, water, sunlight and chlorophyll are requirements for successful photosynthesis.

24.   leaves containing all pigments

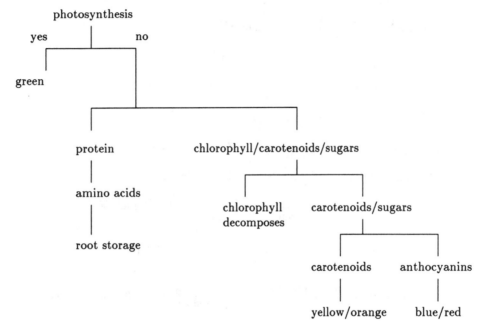

25. 1.   combustion of fossil fuels
     2.   destruction of the rain forests by burning
     3.   increased population

26. Carbon dioxide, methane, and water are all considered to be greenhouse gases. They each act to trap the heat near the surface of the earth in the same manner in which a greenhouse is warmed.

27. The effects of global warming can be reduced by:

     1.   developing new energy sources (not dependent on fossil fuels)
     2.   conservation of energy resources
     3.   recycling
     4.   decreased destruction of the rain forests and other forests

28. About half the carbon dioxide released into the atmosphere remains in the air. The rest is absorbed by plants and used in photosynthesis or is dissolved in the oceans.

CALCULATIONS FROM CHEMICAL EQUATIONS

1. (a) $(25.0 \text{ g KNO}_3)\left(\dfrac{1 \text{ mol}}{101.1 \text{ g}}\right) = 0.247 \text{ mol KNO}_3$

 (b) $(10.8 \text{ g Ca(NO}_3)_2)\left(\dfrac{1 \text{ mol}}{164.1 \text{ g}}\right) = 0.658 \text{ mol Ca(NO}_3)_2$

 (c) $(5.4 \times 10^2 \text{ g (NH}_4)_2\text{C}_2\text{O}_4)\left(\dfrac{1 \text{ mol}}{124.1 \text{ g}}\right) = 4.4 \text{ mol (NH}_4)_2\text{C}_2\text{O}_4$

 (d) $(2.10 \text{ kg NaHCO}_3)\left(\dfrac{1000 \text{ g}}{\text{kg}}\right)\left(\dfrac{1 \text{ mol}}{84.01 \text{ g}}\right) = 25.0 \text{ mol NaHCO}_3$

 (e) $(525 \text{ mg ZnCl}_2)\left(\dfrac{1 \text{ g}}{1000 \text{ mg}}\right)\left(\dfrac{1 \text{ mol}}{136.3 \text{ g}}\right) = 3.85 \times 10^{-3} \text{ mol ZnCl}_2$

 (f) $(56 \text{ mmol NaOH})\left(\dfrac{1 \text{ mol}}{1000 \text{ mmol}}\right) = 0.056 \text{ mol NaOH}$

 (g) $(9.8 \times 10^{24} \text{ molecules CO}_2)\left(\dfrac{1 \text{ mol}}{6.022 \times 10^{23} \text{ molecules}}\right) = 16 \text{ mol CO}_2$

 (h) $(250 \text{ mL C}_2\text{H}_5\text{OH})\left(\dfrac{0.789 \text{ g}}{\text{mL}}\right)\left(\dfrac{1 \text{ mol}}{46.06 \text{ g}}\right) = 4.3 \text{ mol C}_2\text{H}_5\text{OH}$

 (i) $(16.8 \text{ mL solution})\left(\dfrac{1.727 \text{ g}}{\text{mL}}\right)\left(\dfrac{0.800 \text{ g H}_2\text{SO}_4}{\text{g solution}}\right)\left(\dfrac{1 \text{ mol}}{98.08 \text{ g}}\right) = 0.237 \text{ mol H}_2\text{SO}_4$

2. (a) $(2.55 \text{ mol Fe(OH)}_3)\left(\dfrac{106.9 \text{ g}}{\text{mol}}\right) = 273 \text{ g Fe(OH)}_3$

 (b) $(0.00844 \text{ mol NiSO}_4)\left(\dfrac{154.8 \text{ g}}{\text{mol}}\right) = 1.31 \text{ g NiSO}_4$

 (c) $(125 \text{ kg CaCO}_3)\left(\dfrac{1000 \text{ g}}{\text{kg}}\right) = 1.25 \times 10^5 \text{ g CaCO}_3$

 (d) $(0.0600 \text{ mol HC}_2\text{H}_3\text{O}_2)\left(\dfrac{60.05 \text{ g}}{\text{mol}}\right) = 3.60 \text{ g HC}_2\text{H}_3\text{O}_2$

 (e) $(10.5 \text{ mol NH}_3)\left(\dfrac{17.03 \text{ g}}{\text{mol}}\right) = 179 \text{ g NH}_3$

 (f) $(0.725 \text{ mol Bi}_2\text{S}_3)\left(\dfrac{514.2 \text{ g}}{\text{mol}}\right) = 373 \text{ g Bi}_2\text{S}_3$

 (g) $(72 \text{ mmol HCl})\left(\dfrac{1 \text{ mol}}{1000 \text{ mmol}}\right)\left(\dfrac{36.46 \text{ g}}{\text{mol}}\right) = 2.6 \text{ g HCl}$

 (h) $(4.50 \times 10^{21} \text{ molecules C}_6\text{H}_{12}\text{O}_6)\left(\dfrac{1 \text{ mol}}{6.022 \times 10^{23} \text{ molecules}}\right)\left(\dfrac{180.2 \text{ g}}{\text{mol}}\right) = 1.35 \text{ g C}_6\text{H}_{12}\text{O}_6$

 (i) $(500 \text{ mL Br}_2)\left(\dfrac{3.119 \text{ g}}{\text{mL}}\right) = 2 \times 10^3 \text{ g Br}_2$

 (j) $(75 \text{ mL solution})\left(\dfrac{1.175 \text{ g}}{\text{mL}}\right)\left(\dfrac{0.200 \text{ g K}_2\text{CrO}_4}{\text{g solution}}\right) = 18 \text{ g K}_2\text{CrO}_4$

3. **(a)** 10.0 g H_2O or 10.0 g H_2O_2

Water has a lower molar mass than hydrogen peroxide. 10.0 grams of water contains more moles, and therefore more molecules than 10.0 g of H_2O_2.

(b) 25.0 g HCl or 85.0 g $C_6H_{12}O_6$

$$(25.0 \text{ g HCl})\left(\frac{1 \text{ mol}}{36.46 \text{ g}}\right)\left(\frac{6.022 \times 10^{23} \text{ molecules}}{\text{mol}}\right) = 4.12 \times 10^{23} \text{ molecules HCl}$$

$$(85.0 \text{ g } C_6H_{12}O_6)\left(\frac{1 \text{ mol}}{180.2 \text{ g}}\right)\left(\frac{6.022 \times 10^{23} \text{ molecules}}{\text{mol}}\right) = 2.84 \times 10^{23} \text{ molecules } C_6H_{12}O_6$$

HCl contains more molecules

4. Mole ratios

$$2\ C_3H_7OH\ +\ 9\ O_2\ \longrightarrow\ 6\ CO_2\ +\ 8\ H_2O$$

(a) $\dfrac{6 \text{ mol } CO_2}{2 \text{ mol } C_3H_7OH}$

(d) $\dfrac{8 \text{ mol } H_2O}{2 \text{ mol } C_3H_7OH}$

(b) $\dfrac{2 \text{ mol } C_3H_7OH}{9 \text{ mol } O_2}$

(e) $\dfrac{6 \text{ mol } CO_2}{8 \text{ mol } H_2O}$

(c) $\dfrac{9 \text{ mol } O_2}{6 \text{ mol } CO_2}$

(f) $\dfrac{8 \text{ mol } H_2O}{9 \text{ mol } O_2}$

5. Mole ratios

$$3\ CaCl_2\ +\ 2\ H_3PO_4\ \longrightarrow\ Ca_3(PO_4)_2\ +\ 6\ HCl$$

(a) $\dfrac{3 \text{ mol } CaCl_2}{1 \text{ mol } Ca_3(PO_4)_2}$

(d) $\dfrac{1 \text{ mol } Ca_3(PO_4)_2}{2 \text{ mol } H_3PO_4}$

(b) $\dfrac{6 \text{ mol } HCl}{2 \text{ mol } H_3PO_4}$

(e) $\dfrac{6 \text{ mol } HCl}{1 \text{ mol } Ca_3(PO_4)_2}$

(c) $\dfrac{3 \text{ mol } CaCl_2}{2 \text{ mol } H_3PO_4}$

(f) $\dfrac{2 \text{ mol } H_3PO_4}{6 \text{ mol } HCl}$

6. Moles of Cl_2

$$4\ HCl\ +\ O_2\ \longrightarrow\ 2\ Cl_2\ +\ 2\ H_2O$$

$$(5.60 \text{ mol HCl})\left(\frac{2 \text{ mol } Cl_2}{4 \text{ mol HCl}}\right) = 2.80 \text{ mol } Cl_2$$

7. Grams of NaOH

$$Ca(OH)_2\ +\ Na_2CO_3\ \longrightarrow\ 2\ NaOH\ +\ CaCO_3$$

$$(500 \text{ g } Ca(OH)_2)\left(\frac{1 \text{ mol}}{74.10 \text{ g}}\right)\left(\frac{2 \text{ mol NaOH}}{1 \text{ mol } Ca(OH)_2}\right)\left(\frac{40.00 \text{ g}}{\text{mol}}\right) = 5 \times 10^2 \text{ g NaOH}$$

8.

$$Al_4C_3\ +\ 12\ H_2O\ \longrightarrow\ 4\ Al(OH)_3\ +\ 3\ CH_4$$

(a) $(100. \text{ g } Al_4C_3)\left(\dfrac{1 \text{ mol}}{144.0 \text{ g}}\right)\left(\dfrac{12 \text{ mol } H_2O}{1 \text{ mol } Al_4C_3}\right) = 8.33 \text{ mol } H_2O$

(b) $(0.600 \text{ mol } CH_4)\left(\dfrac{4 \text{ mol } Al(OH)_3}{3 \text{ mol } CH_4}\right) = 0.800 \text{ mol } Al(OH)_3$

9. Grams of $Zn_3(PO_4)_2$

$$3 \text{ Zn} + 2 \text{ H}_3PO_4 \longrightarrow Zn_3(PO_4)_2 + 3 \text{ H}_2$$

$$(10.0 \text{ g Zn})\left(\dfrac{1 \text{ mol}}{65.38 \text{ g}}\right)\left(\dfrac{1 \text{ mol } Zn_3(PO_4)_2}{3 \text{ mol Zn}}\right)\left(\dfrac{386.1 \text{ g}}{\text{mol}}\right) = 19.7 \text{ g } Zn_3(PO_4)_2$$

10. $4 \text{ FeS}_2 + 11 \text{ O}_2 \longrightarrow 2 \text{ Fe}_2O_3 + 8 \text{ SO}_2$

(a) $(1.00 \text{ mol } FeS_2)\left(\dfrac{2 \text{ mol } Fe_2O_3}{4 \text{ mol } FeS_2}\right) = 0.500 \text{ mol } Fe_2O_3$

(b) $(4.5 \text{ mol } FeS_2)\left(\dfrac{11 \text{ mol } O_2}{4 \text{ mol } FeS_2}\right) = 12.4 \text{ mol } O_2$

(c) $(1.55 \text{ mol } Fe_2O_3)\left(\dfrac{8 \text{ mol } SO_2}{2 \text{ mol } Fe_2O_3}\right) = 6.20 \text{ mol } SO_2$

(d) $(0.512 \text{ mol } FeS_2)\left(\dfrac{8 \text{ mol } SO_2}{4 \text{ mol } FeS_2}\right)\left(\dfrac{64.06 \text{ g}}{\text{mol}}\right) = 65.6 \text{ g } SO_2$

(e) $(40.6 \text{ g } SO_2)\left(\dfrac{1 \text{ mol}}{64.06 \text{ g}}\right)\left(\dfrac{11 \text{ mol } O_2}{8 \text{ mol } SO_2}\right) = 0.871 \text{ mol } O_2$

(f) $(221 \text{ g } Fe_2O_3)\left(\dfrac{1 \text{ mol}}{159.7 \text{ g}}\right)\left(\dfrac{4 \text{ mol } FeS_2}{2 \text{ mol } Fe_2O_3}\right)\left(\dfrac{120.0 \text{ g}}{\text{mol}}\right) = 332 \text{ g } FeS_2$

11. $MnO_2(s) + 4 \text{ HCl}(aq) \longrightarrow Cl_2(g) + MnCl_2(aq) + 2 \text{ H}_2O$

$(1.05 \text{ mol } MnO_2)\left(\dfrac{4 \text{ mol HCl}}{1 \text{ mol } MnO_2}\right) = 4.20 \text{ mol HCl}$

12.   The balanced equation is $Zn + 2 \text{ HCl} \longrightarrow ZnCl_2 + H_2$

180. g Zn - 35 g Zn = 145 g Zn reacted with HCl

(a) $(145 \text{ g Zn})\left(\dfrac{1 \text{ mol}}{65.38 \text{ g}}\right)\left(\dfrac{1 \text{ mol } H_2}{1 \text{ mol Zn}}\right) = 2.22 \text{ mol } H_2$

(b) $(145 \text{ g Zn})\left(\dfrac{1 \text{ mol}}{65.38 \text{ g}}\right)\left(\dfrac{2 \text{ mol HCl}}{1 \text{ mol Zn}}\right)\left(\dfrac{36.46 \text{ g}}{\text{mol}}\right) = 162 \text{ g HCl}$

13.   The balanced equation is $Fe_2O_3 + 3 \text{ C} \longrightarrow 2 \text{ Fe} + 3 \text{ CO}$

$(125 \text{ kg } Fe_2O_3)\left(\dfrac{1 \text{ kmol}}{159.7 \text{ kg}}\right)\left(\dfrac{2 \text{ kmol Fe}}{1 \text{ kmol } Fe_2O_3}\right)\left(\dfrac{55.85 \text{ kg}}{\text{kmol}}\right) = 87.4 \text{ kg Fe}$

14.   The balanced equation is $3 \text{ Fe} + 4 \text{ H}_2O \longrightarrow Fe_3O_4 + 4 \text{ H}_2$

$(375 \text{ g } Fe_3O_4)\left(\dfrac{1 \text{ mol}}{231.6 \text{ g}}\right)\left(\dfrac{4 \text{ mol } H_2O}{1 \text{ mol } Fe_3O_4}\right)\left(\dfrac{18.02 \text{ g}}{\text{mol}}\right) = 117 \text{ g } H_2O$

$$(375 \text{ g Fe}_3\text{O}_4)\left(\frac{1 \text{ mol}}{231.6 \text{ g}}\right)\left(\frac{3 \text{ mol Fe}}{1 \text{ mol Fe}_3\text{O}_4}\right)\left(\frac{55.85 \text{ g}}{\text{mol}}\right) = 271 \text{ g Fe}$$

15. The balanced equation is $2\ \text{C}_2\text{H}_6 + 7\ \text{O}_2 \longrightarrow 4\ \text{CO}_2 + 6\ \text{H}_2\text{O}$

 (a) $(15.0 \text{ mol C}_2\text{H}_6)\left(\frac{7 \text{ mol O}_2}{2 \text{ mol C}_2\text{H}_6}\right) = 52.5 \text{ mol O}_2$

 (b) $(8.00 \text{ g H}_2\text{O})\left(\frac{1 \text{ mol}}{18.02 \text{ g}}\right)\left(\frac{4 \text{ mol CO}_2}{6 \text{ mol H}_2\text{O}}\right)\left(\frac{44.01 \text{ g}}{\text{mol}}\right) = 13.0 \text{ g CO}_2$

 (c) $(75.0 \text{ g C}_2\text{H}_6)\left(\frac{1 \text{ mol}}{30.07 \text{ g}}\right)\left(\frac{4 \text{ mol CO}_2}{2 \text{ mol C}_2\text{H}_6}\right)\left(\frac{44.01 \text{ g}}{\text{mol}}\right) = 2.20 \times 10^2 \text{ g CO}_2$

16. (a) $\text{KOH} + \text{HNO}_3 \longrightarrow \text{KNO}_3 + \text{H}_2\text{O}$
   16.0 g   12.0 g
   Choose one of the products to calculate. Using KNO_3:

 $$(16.0 \text{ g KOH})\left(\frac{1 \text{ mol}}{56.10 \text{ g}}\right)\left(\frac{1 \text{ mol KNO}_3}{1 \text{ mol KOH}}\right)\left(\frac{101.1 \text{ g}}{\text{mol}}\right) = 28.8 \text{ g KNO}_3$$

 $$(12.0 \text{ g HNO}_3)\left(\frac{1 \text{ mol}}{63.01 \text{ g}}\right)\left(\frac{1 \text{ mol KNO}_3}{1 \text{ mol KOH}}\right)\left(\frac{101.1 \text{ g}}{\text{mol}}\right) = 19.3 \text{ g KNO}_3$$

 Since HNO_3 produces less KNO_3, it is the limiting reactant and KOH is in excess.

 (b) $2\ \text{NaOH} + \text{H}_2\text{SO}_4 \longrightarrow \text{Na}_2\text{SO}_4 + 2\ \text{H}_2\text{O}$
   10.0 g   10.0 g
   Choose one of the products to calculate. Using H_2O:

 $$(10.0 \text{ g NaOH})\left(\frac{1 \text{ mol}}{40.00 \text{ g}}\right)\left(\frac{2 \text{ mol H}_2\text{O}}{2 \text{ mol NaOH}}\right)\left(\frac{18.02 \text{ g}}{\text{mol}}\right) = 4.51 \text{ g H}_2\text{O}$$

 $$(10.0 \text{ g H}_2\text{SO}_4)\left(\frac{1 \text{ mol}}{98.08 \text{ g}}\right)\left(\frac{2 \text{ mol H}_2\text{O}}{1 \text{ mol H}_2\text{SO}_4}\right)\left(\frac{18.02 \text{ g}}{\text{mol}}\right) = 3.67 \text{ g H}_2\text{O}$$

 Since H_2SO_4 produces less H_2O, it is the limiting reactant and NaOH is in excess.

 (c) $2\ \text{Bi(NO}_3)_3 + 3\ \text{H}_2\text{S} \longrightarrow \text{Bi}_2\text{S}_3 + 6\ \text{HNO}_3$
   50.0 g   6.00 g
   Choose one of the products to calculate. Using Bi_2S_3:

 $$(50.0 \text{ g Bi(NO}_3)_3)\left(\frac{1 \text{ mol}}{395.1 \text{ g}}\right)\left(\frac{1 \text{ mol Bi}_2\text{S}_3}{2 \text{ mol Bi(NO}_3)_3}\right)\left(\frac{514.2 \text{ g}}{\text{mol}}\right) = 32.5 \text{ g Bi}_2\text{S}_3$$

 $$(6.00 \text{ g H}_2\text{S})\left(\frac{1 \text{ mol}}{34.08 \text{ g}}\right)\left(\frac{1 \text{ mol Bi}_2\text{S}_3}{3 \text{ mol H}_2\text{S}}\right)\left(\frac{514.2 \text{ g}}{\text{mol}}\right) = 30.2 \text{ g Bi}_2\text{S}_3$$

 Since H_2S produces less Bi_2S_3, it is the limiting reactant and $\text{Bi(NO}_3)_3$ is in excess.

 (d) $3\ \text{Fe} + 4\ \text{H}_2\text{O} \longrightarrow \text{Fe}_3\text{O}_4 + 4\ \text{H}_2$
   40.0 g   16.0 g
   Choose one of the products to calculate. Using H_2:

$$(40.0 \text{ g } Fe)\left(\frac{1 \text{ mol}}{55.85 \text{ g}}\right)\left(\frac{4 \text{ mol } H_2}{3 \text{ mol } Fe}\right)\left(\frac{2.016 \text{ g}}{\text{mol}}\right) = 1.93 \text{ g } H_2$$

$$(16.0 \text{ g } H_2O)\left(\frac{1 \text{ mol}}{18.02 \text{ g}}\right)\left(\frac{4 \text{ mol } H_2}{4 \text{ mol } H_2O}\right)\left(\frac{2.016 \text{ g}}{\text{mol}}\right) = 1.79 \text{ g } H_2$$

Since H_2O produces less H_2, it is the limiting reactant and Fe is in excess.

17. $Fe \quad + \quad CuSO_4 \longrightarrow Cu + FeSO_4$
     2.0 mol $\quad\quad$ 3.0 mol

(a) 2.0 mol Fe react with 2.0 mol $CuSO_4$ to yield 2.0 mol Cu and 2.0 mol $FeSO_4$. 1.0 mol $CuSO_4$ is unreacted. At the completion of the reaction, there will be 2.0 mol Cu, 2.0 mol $FeSO_4$, and 1.0 mol $CuSO_4$.

(b) $$(20.0 \text{ g } Fe)\left(\frac{1 \text{ mol}}{55.85 \text{ g}}\right)\left(\frac{1 \text{ mol } Cu}{1 \text{ mol } Fe}\right)\left(\frac{63.55 \text{ g}}{\text{mol}}\right) = 22.8 \text{ g } Cu$$

$$(40.0 \text{ g } CuSO_4)\left(\frac{1 \text{ mol}}{159.6 \text{ g}}\right)\left(\frac{1 \text{ mol } Cu}{1 \text{ mol } CuSO_4}\right)\left(\frac{63.55 \text{ g}}{\text{mol}}\right) = 15.9 \text{ g } Cu$$

Since $CuSO_4$ produces less Cu, it is the limiting reactant. Determine the mass of $FeSO_4$ produced from $CuSO_4$.

$$(40.0 \text{ g } CuSO_4)\left(\frac{1 \text{ mol}}{159.6 \text{ g}}\right)\left(\frac{1 \text{ mol } FeSO_4}{1 \text{ mol } CuSO_4}\right)\left(\frac{151.9 \text{ g}}{\text{mol}}\right) = 38.1 \text{ g } FeSO_4 \text{ produced}$$

Determine the mass of Fe, which is unreacted.

$$(40.0 \text{ g } CuSO_4)\left(\frac{1 \text{ mol}}{159.6 \text{ g}}\right)\left(\frac{1 \text{ mol } Fe}{1 \text{ mol } CuSO_4}\right)\left(\frac{55.85 \text{ g}}{\text{mol}}\right) = 14.0 \text{ g } Fe \text{ will react}$$

Unreacted Fe = 20.0 g – 14.0 g = 6.0 g. Therefore, at the completion of the reaction, 15.9 g Cu, 38.1 g $FeSO_4$, 6.0 g Fe, and no $CuSO_4$ remain.

18. Limiting reactant calculations

$$C_3H_8 + 5 O_2 \longrightarrow 3 CO_2 + 4 H_2O$$

(a) Reaction between 5.0 mol C_3H_8 and 5.0 mol O_2

$$(5.0 \text{ mol } C_3H_8)\left(\frac{3 \text{ mol } CO_2}{1 \text{ mol } C_3H_8}\right) = 15 \text{ mol } CO_2$$

$$(5.0 \text{ mol } O_2)\left(\frac{3 \text{ mol } CO_2}{5 \text{ mol } O_2}\right) = 3.0 \text{ mol } CO_2$$

The O_2 is the limiting reactant; 3.0 mol CO_2 produced.

(b) Reaction between 3.0 mol C_3H_8 and 20.0 mol O_2

$$(3.0 \text{ mol } C_3H_8)\left(\frac{3 \text{ mol } CO_2}{1 \text{ mol } C_3H_8}\right) = 9.0 \text{ mol } CO_2$$

$$(20.0 \text{ mol } O_2)\left(\frac{3 \text{ mol } CO_2}{5 \text{ mol } O_2}\right) = 12.0 \text{ mol } CO_2$$

The C_3H_8 is the limiting reactant; 9.0 mol CO_2 produced.

(c)   Reaction between 20.0 mol C_3H_8 and 3.0 mol O_2

According to the equation, 1 mol C_3H_8 to 5 mol O_2, O_2 is clearly the limiting reactant.

$$(3.0 \text{ mol } O_2)\left(\frac{3 \text{ mol } CO_2}{5 \text{ mol } O_2}\right) = 1.8 \text{ mol } CO_2 \text{ produced}$$

(d)   Reaction between 2.0 mol C_3H_8 and 14.0 mol O_2

According to the equation, 2 mol C_3H_8 will react with10 mol O_2. Therefore, C_3H_8 is the limiting reactant and 4.0 mol O_2 will remain unreacted.

$$(2.0 \text{ mol } C_3H_8)\left(\frac{3 \text{ mol } CO_2}{1 \text{ mol } C_3H_8}\right) = 6.0 \text{ mol } CO_2 \text{ produced}$$

$$(2.0 \text{ mol } C_3H_8)\left(\frac{4 \text{ mol } H_2O}{1 \text{ mol } C_3H_8}\right) = 8.0 \text{ mol } H_2O \text{ produced}$$

When the reaction is completed, 6.0 mol CO_2, 8.0 mol H_2O, and 4.0 mol O_2 will be in the container.

(e)   Reaction between 20.0 g C_3H_8 and 20.0 g O_2

Convert each amount to grams of CO_2

$$(20.0 \text{ g } C_3H_8)\left(\frac{1 \text{ mol}}{44.09 \text{ g}}\right)\left(\frac{3 \text{ mol } CO_2}{1 \text{ mol } C_3H_8}\right)\left(\frac{44.01 \text{ g}}{\text{mol}}\right) = 59.9 \text{ g } CO_2$$

$$(20.0 \text{ g } O_2)\left(\frac{1 \text{ mol}}{32.00 \text{ g}}\right)\left(\frac{3 \text{ mol } CO_2}{5 \text{ mol } O_2}\right)\left(\frac{44.01 \text{ g}}{\text{mol}}\right) = 16.5 \text{ g } CO_2$$

O_2 is the limiting reactant. The yield is 16.5 g CO_2.

(f)   Reaction between 20.0 g C_3H_8 and 80.0 g O_2

Convert each amount to grams of CO_2

$$(20.0 \text{ g } C_3H_8)\left(\frac{1 \text{ mol}}{44.09 \text{ g}}\right)\left(\frac{3 \text{ mol } CO_2}{1 \text{ mol } C_3H_8}\right)\left(\frac{44.01 \text{ g}}{\text{mol}}\right) = 59.9 \text{ g } CO_2$$

$$(80.0 \text{ g } O_2)\left(\frac{1 \text{ mol}}{32.00 \text{ g}}\right)\left(\frac{3 \text{ mol } CO_2}{5 \text{ mol } O_2}\right)\left(\frac{44.01 \text{ g}}{\text{mol}}\right) = 66.0 \text{ g } CO_2$$

C_3H_8 is the limiting reactant. The yield is 59.9 g CO_2.

(g)   Reaction between 20.0 g C_3H_8 and 200 g O_2

C_3H_8 is the limiting reactant in exercise 18 (f).

Therefore, using 200 g O_2 for the same amount of C_3H_8 will also result in C_3H_8 as the limiting reactant. Thus, the yield will be 59.9 g CO_2.

19. Limiting reactant calculations

$$2 \text{ Al} + 3 \text{ Br}_2 \longrightarrow 2 \text{ AlBr}_3$$

Reaction between 25.0 g Al and 100. g Br_2

$$(25.0 \text{ g Al})\left(\frac{1 \text{ mol}}{26.98 \text{ g}}\right)\left(\frac{2 \text{ mol AlBr}_3}{2 \text{ mol Al}}\right)\left(\frac{266.7 \text{ g}}{\text{mol}}\right) = 247 \text{ g AlBr}_3$$

$(100. \text{ g Br}_2)\left(\dfrac{1 \text{ mol}}{159.8 \text{ g}}\right)\left(\dfrac{2 \text{ mol AlBr}_3}{3 \text{ mol Br}_2}\right)\left(\dfrac{266.7 \text{ g}}{\text{mol}}\right) = 111 \text{ g AlBr}_3$

Br_2 is limiting; 111 g $AlBr_3$ is the theoretical yield of product.

Percent yield $= \left(\dfrac{64.2 \text{ g}}{111 \text{ g}}\right)(100) = 57.8\%$

20. Limiting reactant calculation

$$CO(g) + 2 H_2(g) \longrightarrow CH_3OH(l)$$

Reaction between 40.0 g CO and 10.0 g H_2

$(40.0 \text{ g CO})\left(\dfrac{1 \text{ mol}}{28.01 \text{ g}}\right)\left(\dfrac{1 \text{ mol CH}_3\text{OH}}{1 \text{ mol CO}}\right)\left(\dfrac{32.04 \text{ g}}{\text{mol}}\right) = 45.8 \text{ g CH}_3\text{OH}$

$(10.0 \text{ g H}_2)\left(\dfrac{1 \text{ mol}}{2.016 \text{ g}}\right)\left(\dfrac{1 \text{ mol CH}_3\text{OH}}{2 \text{ mol H}_2}\right)\left(\dfrac{32.04 \text{ g}}{\text{mol}}\right) = 79.5 \text{ g CH}_3\text{OH}$

CO is limiting; H_2 is in excess; 45.7 g CH_3OH will be produced.

$(40.0 \text{ g CO})\left(\dfrac{1 \text{ mol}}{28.01 \text{ g}}\right)\left(\dfrac{2 \text{ mol H}_2}{1 \text{ mol CO}}\right)\left(\dfrac{2.016 \text{ g}}{\text{mol}}\right) = 5.76 \text{ g H}_2 \text{ react}$

$10.0 \text{ g H}_2 - 5.76 \text{ g H}_2 = 4.24 \text{ g H}_2$ remain unreacted

21. Percent yield calculation

$$Fe(s) + CuSO_4(aq) \longrightarrow Cu(s) + FeSO_4(aq)$$

$(400. \text{ g CuSO}_4)\left(\dfrac{1 \text{ mol}}{159.6 \text{ g}}\right)\left(\dfrac{1 \text{ mol Cu}}{1 \text{ mol CuSO}_4}\right)\left(\dfrac{63.55 \text{ g}}{\text{mol}}\right) = 159 \text{ g Cu (theoretical yield)}$

% yield $= \left(\dfrac{\text{actual yield}}{\text{theoretical yield}}\right)(100) = \left(\dfrac{151 \text{ g}}{159 \text{ g}}\right)(100) = 95.0\%$ yield of Cu

22. The balanced equation is $C_6H_{12}O_6 \longrightarrow 2 C_2H_5OH + 2 CO_2$

(a) $(750 \text{ g C}_6\text{H}_{12}\text{O}_6)\left(\dfrac{1 \text{ mol}}{180.2 \text{ g}}\right)\left(\dfrac{2 \text{ mol C}_2\text{H}_5\text{OH}}{1 \text{ mol C}_6\text{H}_{12}\text{O}_6}\right)\left(\dfrac{46.07 \text{ g}}{\text{mol}}\right)(0.846) = 3.2 \times 10^2 \text{ g C}_2\text{H}_5\text{OH}$

(b) $(475 \text{ g C}_2\text{H}_5\text{OH})\left(\dfrac{1 \text{ mol}}{46.07 \text{ g}}\right)\left(\dfrac{1 \text{ mol C}_6\text{H}_{12}\text{O}_6}{2 \text{ mol C}_2\text{H}_5\text{OH}}\right)\left(\dfrac{180.2 \text{ g}}{\text{mol}}\right)\left(\dfrac{1}{0.846}\right) = 1.10 \times 10^3 \text{ g C}_6\text{H}_{12}\text{O}_6$

23. The balanced equation is $3 C + 2 SO_2 \longrightarrow CS_2 + 2 CO_2$

$(950 \text{ g CS}_2)\left(\dfrac{1 \text{ mol}}{76.13 \text{ g}}\right)\left(\dfrac{3 \text{ mol C}}{1 \text{ mol CS}_2}\right)\left(\dfrac{12.01 \text{ g}}{\text{mol}}\right)\left(\dfrac{1}{0.860}\right) = 5.2 \times 10^2 \text{ g C}$

24. The balanced equation is $CaC_2 + 2 H_2O \longrightarrow C_2H_2 + Ca(OH)_2$

$(0.540 \text{ mol C}_2\text{H}_2)\left(\dfrac{1 \text{ mol CaC}_2}{1 \text{ mol C}_2\text{H}_2}\right)\left(\dfrac{64.10 \text{ g}}{\text{mol}}\right) = 34.6 \text{ g CaC}_2$ in the impure sample

$\left(\dfrac{34.6 \text{ g}}{44.5 \text{ g}}\right)(100) = 77.8\% \text{ CaC}_2$ in the impure material

25. The balanced equations are:

$$CaCl_2 \; + \; 2\,AgNO_3 \; \longrightarrow \; Ca(NO_3)_2 \; + \; 2\,AgCl$$

$$MgCl_2 \; + \; 2\,AgNO_3 \; \longrightarrow \; Mg(NO_3)_2 \; + \; 2\,AgCl$$

1 mol of each salt will produce the same amount (2 mol) of AgCl. $MgCl_2$ has a higher percentage of Cl than $CaCl_2$ because Mg has a lower atomic mass than Ca.   Therefore, on an equal mass basis, $MgCl_2$ will produce more AgCl than $CaCl_2$.

Calculations show that 1.00 g $MgCl_2$ produces 3.01 g AgCl, and 1.00 g $CaCl_2$ produces 2.56 g AgCl.

26. The balanced equation is $Li_2O \; + \; H_2O \; \longrightarrow \; 2\,LiOH$

$$\left(\frac{2500 \text{ g } H_2O}{\text{astronaut day}}\right)\left(\frac{1 \text{ mol}}{18.02 \text{ g}}\right)\left(\frac{1 \text{ mol } Li_2O}{1 \text{ mol } H_2O}\right)\left(\frac{29.88 \text{ g}}{\text{mol}}\right)\left(\frac{1 \text{ kg}}{1000 \text{ g}}\right) = 4.1 \;\; \frac{\text{kg } Li_2O}{\text{astronaut day}}$$

$$\left(4.1 \;\; \frac{\text{kg } Li_2O}{\text{astronaut day}}\right)(30 \text{ days})(3 \text{ astronauts}) = 3.7 \times 10^2 \text{ kg } Li_2O$$

27. The balanced equation is

$$H_2SO_4 \; + \; 2\,NaCl \; \longrightarrow \; Na_2SO_4 \; + \; 2\,HCl$$

$$(20.0 \text{ L HCl solution})\left(\frac{1000 \text{ mL}}{1 \text{ L}}\right)\left(\frac{1.20 \text{ g}}{1.00 \text{ mL}}\right)(0.420) \; = \; 9600 \text{ g HCl}$$

$$(9600 \text{ g HCl})\left(\frac{1 \text{ mol}}{36.46 \text{ g}}\right)\left(\frac{1 \text{ mol } H_2SO_4}{2 \text{ mol HCl}}\right)\left(\frac{98.08 \text{ g}}{1 \text{ mol}}\right) \; = \; 13{,}000 \text{ g } H_2SO_4$$

$$(13{,}000 \text{ g } H_2SO_4)\left(\frac{1.00 \text{ g } H_2SO_4 \text{ solution}}{0.96 \text{ g } H_2SO_4}\right)\left(\frac{1\text{kg}}{1000 \text{ g}}\right) = 14 \text{ kg concentrated } H_2SO_4$$

28. The balanced equation is

$$Al(OH)_3{(s)} \; + \; 3\,HCl{(aq)} \; \longrightarrow \; AlCl_3{(aq)} \; + \; 3\,H_2O{(l)}$$

$$(2.5 \text{ L})\left(\frac{3.0 \text{ g HCl}}{L}\right)\left(\frac{1 \text{ mol}}{36.46 \text{ g}}\right)\left(\frac{1 \text{ mol } Al(OH)_3}{3 \text{ mol HCl}}\right)\left(\frac{78.00 \text{ g}}{\text{mol}}\right) = \; 5.3 \text{ g } Al(OH)_3$$

$$(5.3 \text{ g } Al(OH)_3)\left(\frac{1000 \text{ mg}}{\text{g}}\right)\left(\frac{1 \text{ tablet}}{400. \text{ mg}}\right) \; = \; 13 \text{ tablets}$$

29. The balanced equation is $2\,KClO_3 \; \longrightarrow \; 2\,KCl \; + \; 3\,O_2$

12.82 g mixture − 9.45 g residue = 3.37 g O_2 lost by heating

Because the O_2 lost came only from $KClO_3$, we can use it to calculate the amount of $KClO_3$ in the mixture.

$$(3.37 \text{ g } O_2)\left(\frac{1 \text{ mol}}{32.00 \text{ g}}\right)\left(\frac{2 \text{ mol } KClO_3}{3 \text{ mol } O_2}\right)\left(\frac{122.6 \text{ g}}{\text{mol}}\right) \; = \; 8.61 \text{ g } KClO_3$$

$$\left(\frac{8.61 \text{ g } KClO_3}{12.82 \text{ g sample}}\right)(100) \; = \; 67.2\% \;\; KClO_3$$

30. The balanced equation is

$$Ca_3P_2 + 6 H_2O \longrightarrow 3 Ca(OH)_2 + 2 PH_3$$

(a) Correct: $(1 \text{ mol } Ca_3P_2)\left(\dfrac{2 \text{ mol } PH_3}{1 \text{ mol } Ca_3P_2}\right) = 2 \text{ mol } PH_3$

(b) Incorrect: 1 g Ca_3P_2 would produce 0.37 g PH_3

$$(1 \text{ g } Ca_3P_2)\left(\frac{1 \text{ mol}}{182.2 \text{ g}}\right)\left(\frac{2 \text{ mol } PH_3}{1 \text{ mol } Ca_3P_2}\right)\left(\frac{33.99 \text{ g}}{\text{mol}}\right) = 0.37 \text{ g } PH_3$$

(c) Correct: see equation

(d) Correct: see equation

(e) Incorrect: 2 mol Ca_3P_2 requires 12 mol H_2O to produce 4.0 mol PH_3.

$$(2 \text{ mol } Ca_3P_2)\left(\frac{6 \text{ mol } H_2O}{1 \text{ mol } Ca_3P_2}\right) = 12 \text{ mol } H_2O$$

(f) Correct: 2 mol Ca_3P_2 will react with 12 mol H_2O (3 mol H_2O are present in excess) and 6 mol $Ca(OH)_2$ will be formed.

(g) Incorrect: $(200. \text{ g } Ca_3P_2)\left(\frac{1 \text{ mol}}{182.2 \text{ g}}\right)\left(\frac{6 \text{ mol } H_2O}{1 \text{ mol } Ca_3P_2}\right)\left(\frac{18.02}{\text{mol}}\right) = 118 \text{ g } H_2O$

The amount of water present (100. g) is less than needed to react with 200. g Ca_3P_2. H_2O is the limiting reactant.

(h) Incorrect: H_2O is the limiting reactant.

$$(100. \text{ g } H_2O)\left(\frac{1 \text{ mol}}{18.02 \text{ g}}\right)\left(\frac{2 \text{ mol } PH_3}{6 \text{ mol } H_2O}\right)\left(\frac{33.99 \text{ g}}{\text{mol}}\right) = 63.0 \text{ g } PH_3$$

31. The balanced equation is

$$2 CH_4 + 3 O_2 + 2 NH_3 \longrightarrow 2 HCN + 6 H_2O$$

(a) Correct

(b) Incorrect: $(16 \text{ mol } O_2)\left(\frac{2 \text{ mol } HCN}{3 \text{ mol } O_2}\right) = 10.7 \text{ mol } O_2$ (not 12 mol O_2)

(c) Correct

(d) Incorrect: $(12 \text{ mol } HCN)\left(\frac{6 \text{ mol } H_2O}{2 \text{ mol } HCN}\right) = 36 \text{ mol } H_2O$ (not 4 mol H_2O)

(e) Correct

(f) Incorrect: O_2 is the limiting reactant.

$(3 \text{ mol } O_2)\left(\frac{2 \text{ mol } HCN}{3 \text{ mol } O_2}\right) = 2 \text{ mol } HCN$ (not 3 mol HCN)

32. The mask protects the parts of the machine which are not to be etched away.

33. The silicon dioxide is the material which interacts with the chemicals used to etch. The silicon dioxide is removed or sacrificed during this process.

34. Micromachines could be used as smart pills, drug reservoirs, or mini computers. To be effective in these applications, sensing imaging systems and means of locomotion must be devised.

CHAPTER 10

MODERN ATOMIC THEORY

1. An electron orbital is a region in space where an electron is most probably found.

2. A second electron may enter an orbital already occupied by an electron if its spin is opposite the electron already in the orbital and all other orbitals of the same sublevel contain an electron.

3. All the electrons in the atom are located in the orbitals closest to the nucleus.

4. Both 1s and 2s orbitals are spherical in shape and located symmetrically around the nucleus. The sizes of the spheres are different - the radius of 2s is larger than the 1s. The electrons in 2s orbitals are located further from the nucleus.

5. The energy sublevels are s, p, d, and f.

6. 1s, 2s, 2p, 3s, 3p, 4s, 3d, 4p

7. s - 2 per shell
    p - 6 per shell after the first level
    d - 10 per shell after the second level

8. | | | | | |
    |---|---|---|---|---|
    | H | 1 proton | | U | 92 protons |
    | B | 5 protons | | Br | 35 protons |
    | F | 9 protons | | Sb | 51 protons |
    | Sc | 21 protons | | Pb | 82 protons |
    | Ag | 47 protons | | | |

9. B $1s^2 2s^2 2p^1$

    Ti $1s^2 2s^2 2p^6 3s^2 3p^6 4s^2 3d^2$

    Zn $1s^2 2s^2 2p^6 3s^2 3p^6 4s^2 3d^{10}$

    Br $1s^2 2s^2 2p^6 3s^2 3p^6 4s^2 3d^{10} 4p^5$

    Sr $1s^2 2s^2 2p^6 3s^2 3p^6 4s^2 3d^{10} 4p^6 5s^2$

10. The eleventh electron of sodium is located in the third energy level because the first and second levels are filled.

11. The main difference is that the Bohr orbit has an electron traveling a specific path while an orbital is a region in space where the electron is most probably found.

12. The spectral lines of hydrogen are produced by the release of energy emitted when an electron falls from a higher energy level to a lower energy level (closer to the nucleus).

13. Bohr said that a number of orbits were available, each corresponding to an energy level. When an electron falls from a higher energy orbit to a lower orbit energy is given off as a specific frequency of light. Only those energies are seen and form the hydrogen spectrum. Each line corresponds to a change from one orbit to another.

14. Bohr's model was inadequate since it could not account for atoms more complex than hydrogen. It was modified by Schrodinger into the modern concept of the atom in which electrons exhibit wave and particle properties. The motion of electrons is determined only by probability functions as a region in space, or a cloud surrounding the nucleus.

15.

$\underline{n}$	$\underline{2n^2}$
1	2
2	8
3	18
4	32
5	50
6	72

16. 32 electrons in the fourth energy level: $2(4)^2 = 32$

17. 9 orbitals in the third energy level: $3s$, $3p_x$, $3p_y$, $3p_z$ plus five d orbitals

18. See Figure 10.2 in Section 10.4

19.

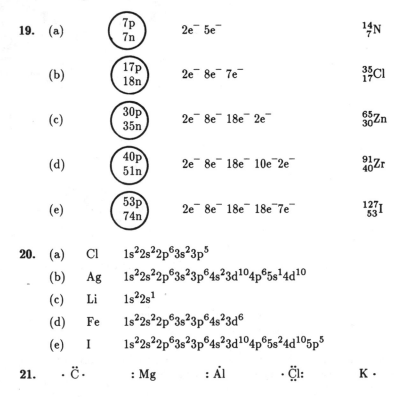

(a) 7p 7n $2e^-$ $5e^-$ $^{14}_{7}N$

(b) 17p 18n $2e^-$ $8e^-$ $7e^-$ $^{35}_{17}Cl$

(c) 30p 35n $2e^-$ $8e^-$ $18e^-$ $2e^-$ $^{65}_{30}Zn$

(d) 40p 51n $2e^-$ $8e^-$ $18e^-$ $10e^-$ $2e^-$ $^{91}_{40}Zr$

(e) 53p 74n $2e^-$ $8e^-$ $18e^-$ $18e^-$ $7e^-$ $^{127}_{53}I$

20. (a) Cl $\quad 1s^2 2s^2 2p^6 3s^2 3p^5$

(b) Ag $\quad 1s^2 2s^2 2p^6 3s^2 3p^6 4s^2 3d^{10} 4p^6 5s^1 4d^{10}$

(c) Li $\quad 1s^2 2s^1$

(d) Fe $\quad 1s^2 2s^2 2p^6 3s^2 3p^6 4s^2 3d^6$

(e) I $\quad 1s^2 2s^2 2p^6 3s^2 3p^6 4s^2 3d^{10} 4p^6 5s^2 4d^{10} 5p^5$

21. $\cdot \ddot{C} \cdot$ $\qquad$: Mg $\qquad$: $\dot{A}l$ $\qquad$ $\cdot \ddot{\underset{..}{C}l}$: $\qquad$ K $\cdot$

22. 3 is the third energy level
d indicates an energy sublevel
7 indicates the number of electrons in the d sublevel

23.

		1s	2s	2p	3s	3p	4s	3d

(a) Si [↑↓] [↑↓] [↑↓][↑↓][↑↓] [↑↓] [↑][↑][ ]

(b) S [↑↓] [↑↓] [↑↓][↑↓][↑↓] [↑↓] [↑↓][↑][↑]

(c) Ar [↑↓] [↑↓] [↑↓][↑↓][↑↓] [↑↓] [↑↓][↑↓][↑↓]

(d) V [↑↓] [↑↓] [↑↓][↑↓][↑↓] [↑↓] [↑↓][↑↓][↑↓] [↑↓] [↑][↑][↑][ ][ ]

24. Noble gases all have filled s and p orbitals in the outermost energy level.

25. The last electron in potassium is located in the fourth energy level because the 4s orbital is at a lower energy level than the 3d orbital.

26. (a) Mg (b) F (c) Ni (d) Mn (e) Mo

27.

	atomic #	electron structure
(a)	8	$1s^2 2s^2 2p^4$
(b)	11	$1s^2 2^2 2p^6 3s^1$
(c)	17	$1s^2 2s^2 2p^6 3s^2 3p^5$
(d)	23	$1s^2 2s^2 2p^6 3s^2 3p^6 4s^2 3d^3$
(e)	28	$1s^2 2s^2 2p^6 3s^2 3p^6 4s^2 3d^8$
(f)	34	$1s^2 2s^2 2p^6 3s^2 3p^6 4s^2 3d^{10} 4p^4$

28.

	atomic #	electron structure
(a)	9	$1s^2 2s^2 2p^5$
(b)	26	$[Ar]4s^2 3d^6$
(c)	31	$[Ar]4s^2 3d^{10} 4p^1$
(d)	39	$[Kr]5s^2 4d^1$
(e)	52	$[Kr]5s^2 4d^{10} 5p^4$

29. (a) Sc (b) Kr (c) Sn (d) Cs

30. (a) S (b) Ni

31. In a scanning-probe microscope, a probe is placed near the surface of a sample and a parameter such as voltage is measured. The signals are translated electronically into a topographic image of the object. In a light microscope the image is formed by light reflecting from the object.

32. The object to be viewed is coated with a thin coating of metal and a probe is placed near the surface. Electrons jump from the metal to the probe producing the image by generating a varying current.

33. In the atomic force microscope the probe measures the electric forces between electrons in the molecule, while in scanning-tunneling the actual movement of electrons is measured.

34. (a) $\left(\begin{smallmatrix} 14n \\ 13p \end{smallmatrix}\right)$ $2e^-\ 8e^-\ 3e^-$ $^{27}_{13}Al$

   (b) $\left(\begin{smallmatrix} 28n \\ 23p \end{smallmatrix}\right)$ $2e^-\ 8e^-\ 11e^-\ 2e^-$ $^{51}_{23}V$

35. Many elements tend to attain an electron structure of eight electrons in the outermost energy level during chemical changes. This structure is identical to that of the noble gases (except **He**) and has great stability associated with it.

36.   : He   : B ·   · Ö ·   Na ·   · Si ·

   : Är: · Ga   Ca: : Br ·   : Kr:

37. The correct statements are a, c, d, e, f, h, i, j, k, l, n, p, q, r, u.

   (b) The maximum number of electrons in the first energy level is 2.

   (g) The electron structure of a calcium atom is $1s^2 2s^2 2p^6 3s^2 3p^6 4s^2$.

   (m) The Lewis dot symbol for potassium is K ·

   (o) A p orbital is shaped like a dumbell on either side of the nucleus.

   (s) The Lewis dot symbol for the noble gas helium is He:

   (t) When an orbital contains two electrons, they have opposite spins.

   (v) Rutherford concluded from his experiment that the positive charge and almost all the mass were concentrated in a very small nucleus.

CHAPTER 11

THE PERIODIC TABLE

1. All elements in the s block have their outermost electron(s) in an s orbital.

2. | Atomic # | Symbol |
    |----------|--------|
    | 8 | O |
    | 16 | S |
    | 34 | Se |
    | 52 | Te |
    | 84 | Po |

    All these elements have an outermost electron structure of s^2p^4.

3. (a) germanium

    (b) predicted density 5.5 g/ mL, actual density 5.35 g/mL

4. The size of the elements in the 3rd period decreases from left to right.

5. F, Cl, Br, I, At

6. Li, Na, K, Rb, Cs, Fr

7. The greatest number of elements in any period is 32. The 6th period has this number.

8. The elements in Group A always have their last electrons in the outermost energy level, while those in Group B lie in an inner level.

9. Dobereiner's triads were elements in groups of three. The three elements in each triad were similar in properties with the center element, reflecting an average of the other two in some respects. These observations indicated a recurrance of properties among some elements, and pointed towards periodicity among the elements.

10. Newlands' Law of Octaves is based upon the evidence that up to element 21, where the d electrons first appear, properties do repeat with each eighth element. This is because an s orbital contains 2 electrons and three p orbitals contain 6 electrons. Thus, for every 8 electrons, the element will have the same number of valence electrons and result in similar properties. Newlands' law repeated every seventh element, like musical scales, because the noble gases were not known at the time.

11. Newlands' Law of Octaves does not extend as far as bromine. The appearance of d electrons, beginning with Sc, element 21, disrupts the sequence of eight.

12. Periodicity indicates that as the elements are arranged consecutively in order of increasing atomic number, a periodic recurrence of similarities in the properties of the elements is observed.

13. Vacancies in Mendeleev's periodic table indicated the existence of elements not yet discovered at that time. From the position of the vacancy in the periodic table, expected properties of the missing element could be proposed.

14. Three major changes have been made to the periodic table since Mendeleev's time: (1) a new family of elements, the noble gases, was discovered and added; (2) the elements are arranged by atomic number, not atomic mass; (3) elements having atomic numbers greater than 92 have been synthesized and fitted into the table.

15. Mosley's efforts indicated the periodic arrangement of the elements be based on the atomic numbers of the elements rather than on the atomic masses of the elements.

16. More energy is required for neon because it has a very stable outer electron structure consisting of an octet of electrons in filled orbitals (noble gas electron structure). Sodium, an alkali metal, has a relatively unstable outer electron structure with a single electron in an unfilled orbital. The sodium electron is also farther away from the nucleus and is shielded by more inner electron shells than are neon outer shell electrons.

17. When a third electron is removed from beryllium, it must come from a very stable electron shell structure corresponding to that of the noble gas, helium. In addition, the third electron must be removed from a +2 beryllium ion, which increases the difficulty of removing it.

18. The first ionization energy decreases from top to bottom because in the successive alkali metals, the outermost electron is farther away from the nucleus and is shielded from the positive nucleus by additional electron shells.

19. The first ionization energy decreases from top to bottom because the outermost electrons in the successive noble gases are farther away from the nucleus and are shielded by additional inner electron shells.

20. Barium and beryllium are in the same family. The electron to be removed from barium is, however, located in an energy level father away from the nucleus than is the energy level holding the electron in beryllium. Hence, it requires less energy to remove the electron from barium than to remove the electron from beryllium. Barium, therefore, has a lower ionization energy than beryllium.

21. The first electron removed from a sodium atom is the one outer-shell electron, which is shielded from most of its nuclear charge by the electrons of lower levels. To remove a second electron from the sodium ion requires breaking into the noble gas structure. This requires much more energy than that necessary to remove the first electron, because the Na^+ is already positive.

22. Pairs of elements which are out of sequence with respect to atomic masses are: Ar and K; Co and Ni; Te and I; Th and Pa; U and Np; Es and Fm; Md and No.

23. Moving from left to right in any period of elements, the atomic number increases by one from one element to the next and the atomic radius generally decreases. Each period (except period 1) begins with an alkali metal and ends with a noble gas. There is a trend in properties of the elements changing from metallic to nonmetallic from the beginning to the end of the period.

24. All the elements in a Group have the same number of outer shell electrons.

25. All the alkali metals have one electron in their outer energy level.

26. Zinc, cadmium, and mercury have two electrons in their outer energy level. Each of the three has eighteen electrons in their next to outermost energy level.

27. Electron dot formulas: Cs ·   Ba: Ṫl: · Ṗb: · P̤o: · Ä̤t: :R̤n:

Each of these is a representative element and has the same number of electrons in its outer shell as its periodic group number.

28. Elements in the same family of elements:

(a) and (g)	Group IA.
(b) and (d)	Group VIA.

29. Elements in the same family

(a) and (f)	noble gases
(e) and (h)	IB

30. 12 (Mg) and 38 (Sr) have properties most resembling Ca, as they are in the same family with the same outer shell electron structure.

31. Oxygen and sulfur are in the same family because each has six electrons in the outermost energy level.

32.

(a)	K	metal	(e)	I	nonmetal
(b)	Pu	metal	(f)	W	metal
(c)	S	nonmetal	(g)	Mo	metal
(d)	Sb	metalloid	(h)	Ge	metalloid

33. An electron first appears in a d orbital in period 4, Group IIIB.

34. Group IIIA elements contain 3 electrons in the outer shell.
Group IIIB elements contain 2 electrons in the outer shell and one electron in an inner d shell.
Group A elements are representative, while Group B elements are transition elements.

35. Transition elements are found in IB, IIB, IIIB, IVB, VB, VIB, VIIB, and VIII, or 3, 4, 5, 6, 7, 8, 9, 10, 11, 12.

36. In transition elements the last electron added is in a d or f orbital. The last electron added in a representative element is an s or p electron.

37.

(a)	K > Na	(d)	I > Br
(b)	Na > Mg	(e)	Zr > Ti
(c)	O > F		

38. The first element in each group has the smallest radius.

39. Atomic size increases down a column since each successive element has an additional energy level which contains electrons located farther from the nucleus.

40.

Group	IA	IIA	IIIA	IVA	VA	VIA	VIIA
	E ·	E :	Ė:	· Ė:	· Ë:	· Ë:	: Ë:

41. The discovery of isotopes was evidence that atomic mass was not a characteristic property for elements. Instead, the atomic number was verified as a fundamental characteristic for each element. The atomic mass can vary (isotopes) without changing elements. When the atomic number is changed, the element differs.

42. 8, 16, 34, 52, 84. All have 6 electrons in their outer shell.

43. The outermost energy level is the 7th. This element contains one electron in the 7s orbital.

44. If 36 is a noble gas, 35 would be in VIIA and 37 would be in IA.

45. Answers will vary.

46. $Cl < Mg < Na < K < Rb$ with regard to atomic radius.

47. The correct statements are a, d, e, g, h, i, k, n, o, p, t, v.

(b)   There are fewer nonmetallic elements than metallic elements

(c)   Metallic properties of the elements decrease from left to right across a period.

(f)   Potassium belongs to the alkali metal family.

(j)   An atom of oxygen is smaller than an atom of lithium.

(l)   An atom of aluminum (Group IIIA) has three electrons in its outer shell.

(m)   Elements with low ionization energy tend to have metallic properties.

(q)   The element [Kr] $5s^2$ is a metal.

(r)   The element with Z = 12 forms compounds similar to the element with Z = 38.

(s)   Nitrogen, fluorine, neon, and bromine are all nonmetals.

(u)   The atom having an outer shell electron structure of $5s^2 5p^2$ would be in period 5, Group IVA.

(w)   The most metallic element of Group VIIA is astatine.

48. A superconductor is a substance in which the electrical resistance drops to zero at very low temperatures.

49. 77 K is the boiling point of liquid N_2. If a substance superconducts at this temperature, liquid N_2 can be used to cool the material. Since liquid N_2 is relatively inexpensive, superconductivity at this temperature would be more economical.

50. Chu realized that elements in the same family have similar chemistry. By substituting an element in the same family with a smaller radius, he brought the layers closer together without changing the chemistry. This raised the temperature for superconductivity.

51. Current material used in superconductors is brittle, non-malleable and does not carry a high current per cross sectional area.

CHEMICAL BONDS:
THE FORMATION OF COMPOUNDS FROM ATOMS

1. (a) Elements with the highest electronegativities are found in the upper right hand corner of the periodic table.

   (b) Elements with the lowest electonegativities are found in the lower left of the periodic table.

2. Magnesium atom is larger, since it has electrons in the 3rd shell, while a magnesium ion does not. Also, the ion has 12 protons and 10 electrons, creating a charge imbalance and drawing the electrons in towards the nucleus more closely.

3. A bromine atom is smaller since it has one less electron than the bromine ion in its outer shell. Also, the ion has 35 protons and 36 electrons, creating a charge imbalance, resulting in a lessening of the attraction of the electrons towards the nucleus.

4.

	+	−		+	−		+	−
(a)	H	O	(e)	N	O	(i)	C	Cl
(b)	Na	F	(f)	H	C	(j)	I	Br
(c)	H	N	(g)	H	Cl	(k)	Mg	H
(d)	Pb	S	(h)	Li	H	(l)	O	F

5.

(a)	ionic	(e)	covalent
(b)	covalent	(f)	ionic
(c)	covalent	(g)	covalent
(d)	ionic	(h)	ionic

6. While the number of electrons in the outermost energy level is the same for all elements in a family, the energy level in which they are located is not always the same. As one travels down a family of elements, the valence electrons are successively located in energy levels farther away from the nucleus. The farther the energy level is from the nucleus, the greater will be the energy of the electrons in that energy level and the less will be the additional energy required to remove the electron from the atom.

   Magnesium has an electron structure $1s^2 2s^2 2p^6 3s^2$, while the structure for chlorine is $1s^2 2s^2 2p^6 3s^2 3p^5$. When these two elements react with each other, each magnesium atom loses its $3s^2$ electrons, one to each of two chlorine atoms. The resulting structures for both magnesium and chlorine are noble gas configurations.

7. (a) $F + 1e^- \longrightarrow F^-$ (b) $Ca \longrightarrow Ca^{2+} + 2e^-$

8. (a) $Mg: + \cdot \ddot{F}: + \cdot \ddot{F}: \longrightarrow MgF_2$ (c) $Ca: + \cdot \ddot{O}: \longrightarrow CaO$

   (b) $K\cdot + \cdot K + \cdot \ddot{O}: \longrightarrow K_2O$ (d) $Na\cdot + \cdot \ddot{B}r: \longrightarrow NaBr$

CHAPTER 12

CHEMICAL BONDS
THE FORMATION OF COMPOUNDS FROM ATOMS

1. (a) Elements with the highest electronegativities are found in
        the upper right hand corner of the periodic table.

   (b) Elements with the lowest electonegativities are found in the
        lower left of the periodic table.

2. Magnesium atom is larger because it has electrons in the 3rd shell, while a magnesium ion does
   not. Also, the ion has 12 protons and 10 electrons, creating a charge imbalance and drawing
   the electrons in towards the nucleus more closely.

3. A bromine atom is smaller because it has one less electron than the bromine ion in its outer
   shell. Also, the ion has 35 protons and 36 electrons, creating a charge imbalance, resulting in a
   lessening of the attraction of the electrons towards the nucleus.

4.   +   −   +   −   +   −

   (a) H   O   (e)   N   O   (i)   C   Cl
   (b) Na   F   (f)   H   C   (j)   I   Br
   (c) H   N   (g)   H   Cl   (k)   Mg   H
   (d) Pb   S   (h)   Li   H   (l)   O   F

5. (a) ionic   (e)   covalent
   (b) covalent   (f)   ionic
   (c) covalent   (g)   covalent
   (d) ionic   (h)   ionic

6. Magnesium has an electron structure $1s^2 2s^2 2p^6 3s^2$, while·the structure for chlorine is
   $1s^2 2s^2 2p^6 3s^2 3p^5$.   When these two elements react with each other, each magnesium atom loses
   its $3s^2$ electrons, one to each of two chlorine atoms.   The resulting structures for both
   magnesium and chlorine are noble gas configurations.

7. (a) $F + 1e^- \longrightarrow F^-$ (b) $Ca \longrightarrow Ca^{2+} + 2e^-$

8. (a) $Mg\colon + \cdot\ddot{F}\colon + \cdot\ddot{F}\colon \longrightarrow MgF_2$ (c) $Ca\colon + \cdot\ddot{O}\colon \longrightarrow CaO$

   (b) $K\cdot + \colon K + \cdot\ddot{O}\colon \longrightarrow K_2O$ (d) $Na\cdot + \cdot\ddot{Br}\colon \longrightarrow NaBr$

9. Valence electrons are the electrons found in the outermost energy level of an atom.

(d)　　· Ga⁝　　　　　(e)　　Ga^{3+}

21. Sodium chloride exists as aggregates of sodium and chloride ions. An electron is transferred from a sodium atom to a chlorine atom forming these ions. Substances with ionic bonds do not exist as molecules. The ions are held together by ionic bonds.

22. A covalent bond results from the sharing of a pair of electrons between two bonding atoms; whereas, an ionic bond involves the transfer of one electron or more from one atom to another.

23. In a coordinate covalent bond, the electrons being shared are donated by one of the atoms; whereas, in an ordinary covalent bond, electrons shared come from both atoms.

24. (a) covalent　　　　(e) covalent
    (b) ionic　　　　　(f) ionic
    (c) ionic　　　　　(g) covalent
    (d) covalent　　　　(h) covalent

25. (a) ionic　　　　　(d) covalent
    (b) covalent　　　　(e) covalent
    (c) covalent

26. (a) H ⁚ H　　　　(b) :N ⦂⦂⦂ N:　　　(c) :C̈l ⁚ C̈l:

    (d) : Ö ⁚⁚ Ö:　　　(e) :B̈r ⁚ B̈r:

27. (a)  :C̈l ⁚ N̈ ⁚ C̈l:　　　(b) H ⁚ Ö ⁚ C ⁚⁚ Ö:
    　　　　:C̈l:　　　　　　　　　　　　:Ö:
    　　　　　　　　　　　　　　　　　　　H

    　　　　:Ö:　　　　　　　　　H　H
    (c) H ⁚ Ö ⁚P̈⁚ Ö ⁚ H　　(d) 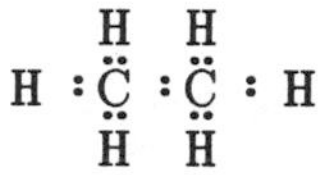 H ⁚C̈ ⁚C̈ ⁚ H
    　　　　:Ö:　　　　　　　　　H　H

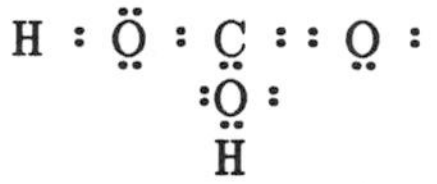

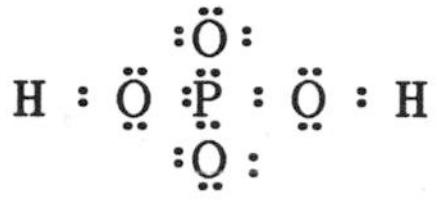

    (e) 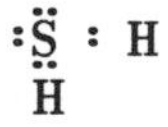 :S̈ ⁚ H　　　　　(f) 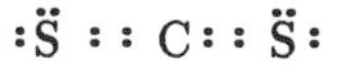 :S̈ ⁚⁚ C ⁚⁚ S̈:
    　　　H

28. The structure is drawn to indicate covalent bonds. Since the electronegativity difference between Na and O is > 1.7, the compound is ionic. A better representation would be
    $[Na]^{+}[:Ö:]^{2-}[Na]^{+}$

29. The four most electronegative elements are F, O, N, Cl.

30. (a) $[Ba]^{2+}$

    (b) $[Al]^{3+}$

    (c) $[:Ï:]^{1-}$

    (d) $[:S̈:]^{2-}$

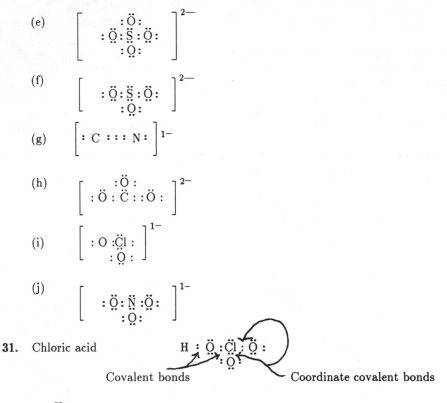

(e)
$$\left[\begin{array}{c} :\ddot{O}: \\ :\ddot{O}:\ddot{S}:\ddot{O}: \\ :\ddot{O}: \end{array} \right]^{2-}$$

(f)
$$\left[\begin{array}{c} :\ddot{O}:\ddot{S}:\ddot{O}: \\ :\ddot{O}: \end{array} \right]^{2-}$$

(g)
$$\left[:C:::N: \right]^{1-}$$

(h)
$$\left[\begin{array}{c} :\ddot{O}: \\ :\ddot{O}:\ddot{C}::\ddot{O}: \end{array} \right]^{2-}$$

(i)
$$\left[\begin{array}{c} :O:\ddot{C}l: \\ :\ddot{O}: \end{array} \right]^{1-}$$

(j)
$$\left[\begin{array}{c} :\ddot{O}:\ddot{N}:\ddot{O}: \\ :\ddot{O}: \end{array} \right]^{1-}$$

31. Chloric acid

$$H:\ddot{O}:\ddot{C}l:\ddot{O}: \\ :\ddot{O}:$$

Covalent bonds   Coordinate covalent bonds

32.
$$H:\overset{\displaystyle H}{\underset{\displaystyle \cdot\cdot}{N}}:H$$

All bonds are covalent. Ammonia can form a coordinate covalent bond by using the unbonded pair of electrons on the nitrogen.

33. high ⟵ Electronegativity ⟶ low

   F > O > S > H > Mg > Cs

34. (a) polar   (d) nonpolar
    (b) polar   (e) nonpolar
    (c) nonpolar   (f) polar

35. It is possible for a molecule to be nonpolar even though it contains polar bonds. If the molecule is symmetrical, the polarities of the bonds will cancel (in a manner similar to a positive and negative number of the same size) resulting in a nonpolar molecule.

36. Both molecules contain polar bonds. CO_2 is symmetrical about the C atom, so the polarities cancel. In CO, there is only one polar bond, therefore the molecule is polar.

37. The correct statements are b, c, e, h, i, j, l, o, q, r, s, t, u, v, w, x, z, bb, cc, gg, hh, jj, kk, mm, nn.

(a) If the formula for calcium iodide id CaI_2, then the formula for cesium iodide is CsI.

(d) Sodium and chlorine do not form molecules of NaCl.

(f) The noble gases have no tendancy to lose electrons.

(g) The chemical bonds in a water molecule are covalent.

(k) Fluorine has the highest electronegativity of all the elements.

(m) A cation is smaller than its corresponding neutral atom.

(n) Cl_2 is more covalent in character than HCl.

(p) A nitrogen atom has five valence electrons.

(y) The simplest compound between oxygen and fluorine F_2O.

(aa) The smaller the difference in electronegativity between two atoms, the more covalent the bond between them will be.

(dd) The correct Lewis structure for CO_2 is : Ö :: C :: Ö:

(ee) The correct Lewis structure for $SO_4{}^{2-}$ is
$$\left[\begin{array}{c} :\ddot{O}: \\ :\ddot{O}:S:\ddot{O}: \\ :\ddot{O}: \end{array} \right]^{2-}$$

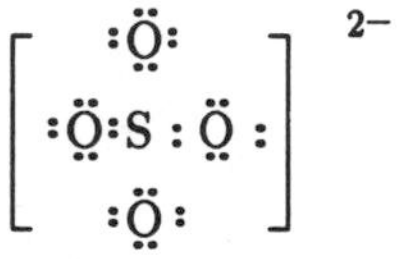

(ff) In period 5, the element having the highest ionization energy is Xe.

(ii) The structures that show H_2O as a dipole and CO_2 as not a dipole are

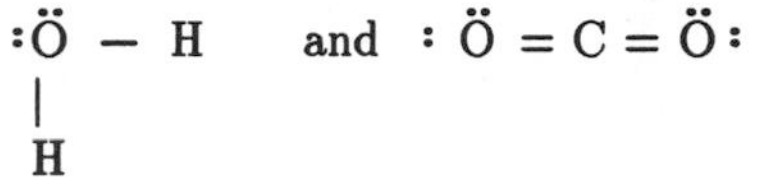

:Ö — H   and : Ö = C = Ö :
    |
    H

(ll) A molecule with a central atom having 3 bonding and one nonbonding electron pair will have a tetrahedral electron structure and a pyrasmidal shape.

38. (a) tetrahedral
    (b) pyramidal
    (c) bent

39. (a) 4 pairs e^-, linear
    (b) 4 pairs e^-, bent
    (c) 3 pairs e^-, trigonal planar

40. (a) 105°
    (b) 107°
    (c) 109°
    (d) 109°

41. (a) square planar
    (b) pyramidal
    (c) tetrahedral

42. Liquid crystals are used in mood rings, calculators, watches, thermometers, bullet resistant vests, canoes and the space shuttle.

43. Normally, the LCD acts as a mirror reflecting light. It has a series of layers, however. When the molecules at the top align with the lines etched on the first layer of glass and those at the bottom align with the grooves on the bottom layer of glass, the molecules in between form a twisted spiral (trying to align with those near them). If a current is applied to specific segments of the etched glass, the plates become charged and the spirals of molecules are attached to the charged plate destroying the arrangement. The pattern of reflected light is changed and a number appears.

44. Kevlar is a nematic liquid crystal which contains molecules all pointing in the same direction. It gets super strength by passing through a liquid crystal state in the manufacturing process. During this time the molecules are aligned and formed into fibers.

CHAPTER 13

THE GASEOUS STATE OF MATTER

1. In Figure 13.1, color is the evidence of diffusion; bromine is colored and air is colorless. If hydrogen and oxygen had been the two gases, this would not work because both gases are colorless. Two ways could be used to show the diffusion. The change of density would be one method. Before diffusion the hydrogen bulb would be much less dense. After diffusion the densities would be equal. A second method would require the introduction of spark gaps into both flasks. Before diffusion, neither gas would show a reaction when sparked. After diffusion, the gases in both flasks would explode because of the mixture of the two gases.

2. The air pressure inside the balloon is greater than the air pressure outside the balloon. The pressure inside must equal the sum of the outside air pressure plus the pressure exerted by the stretched rubber of the balloon.

3. The major components of dry air are nitrogen and oxygen.

4. 1 torr = 1 mm Hg

5. The molecules of H_2 at 100°C are moving faster. Temperature is a measure of average kinetic energy. At higher temperatures, the molecules will have more kinetic energy.

6. 1 atm corresponds to 4 L.

7. The pressure times the volume at any point on the curve is equal to the same value. This is an inverse relationship as is Boyle's law. (PV = k)

8. If $T_2 < T_1$, the volume of the cylinder would decrease. (the piston would move downward)

9. The pressure inside the bottle is less than atmospheric pressure. This can be concluded since the water inside the bottle is higher than the water in the trough.

10. The density of air is given as 1.29 g/L. Any gas with a greater density is listed below air on the table. Any five of these gases would be correct.

11. Basic assumptions of Kinetic Molecular Theory include:

    (a)   Gases consist of tiny molecules.

    (b)   The distance between molecules is great compared to the size of the molecules.

    (c)   Gas molecules move in straight lines. They collide with one another and with the walls of the container with no loss of energy.

    (d)   Gas molecules have no attraction for each other.

    (e)   The average kinetic energy of all gases is the same at any given temperature. It varies directly with temperature.

12. The order of increasing molecular velocities is the order of decreasing molar masses.

molecular velocity increases $\longrightarrow$

$Rn,\ F_2,\ N_2,\ CH_4,\ He,\ H_2$

$\longleftarrow$ molar mass increases

At the same temperature the kinetic energies of the gases are the same and equal to $\frac{1}{2}mv^2$. For the kinetic energies to be the same, the velocities must increase as the molar masses decrease.

13. Average kinetic energies of all these gases are the same as the samples are all at the same temperature.

14. Gases are described by the following parameters:

(a) Pressure (c) Temperature
(b) Volume (d) Number of moles

15. An ideal gas is one which follows the gas laws at all P, V, and T and whose behavior is described exactly by the Kinetic Molecular Theory.

16. A gas is least likely to behave ideally at low temperatures. Under this condition, the velocities of the molecules decrease and attractive forces between the molecules begin to play a significant role.

17. A gas is least likely to behave ideally at high pressures. Under this condition, the molecules are close enough to each other so that their volume is no longer small compared to the volume of the container. Attractive forces may also occur here and sooner or later, the gas will liquefy.

18. Equal volumes of H_2 and O_2 at the same T and P:

(a) have equal numbers of molecules (Avogadro's law)

(b) mass O_2 = 16 (mass H_2) 16 × mass H_2

(c) moles O_2 = moles H_2

(d) average kinetic energies are the same (T same)

(e) rate H_2 = 4 (rate O_2) (Graham's Law of Effusion)

(f) density O_2 = 16 (density H_2)

$$\text{density } O_2 = \left(\frac{\text{mass } O_2}{\text{volume } O_2}\right) \qquad \text{density } H_2 = \left(\frac{\text{mass } H_2}{\text{volume } H_2}\right)$$

$$\text{volume } O_2 = \text{volume } H_2$$

$$\left(\frac{\text{mass } O_2}{\text{den } O_2}\right) = \left(\frac{\text{mass } H_2}{\text{den } H_2}\right) \qquad \text{density } O_2 = \left(\frac{\text{mass } O_2}{\text{mass } H_2}\right)(\text{den } H_2)$$

$$= \left(\frac{16}{1}\right)(\text{den } H_2)$$

19. Behavior of gases as described by the Kinetic Molecular Theory.

(a) Boyle's law. Boyle's law states that the volume of a fixed mass of gas is inversely proportional to the pressure, at constant temperature. The Kinetic Molecular Theory assumes the volume occupied by gases is mostly empty space. Decreasing the volume of a gas by compressing it, increases the concentration of gas molecules, resulting in more collisions of the molecules and increased pressure upon the walls of the container.

(b) Charles' law. Charles' law states that the volume of a fixed mass of gas is directly proportional to the absolute temperature, at constant pressure. According to Kinetic Molecular Theory, the kinetic energies of gas molecules are proportional to the absolute temperature. Increasing the temperature of a gas causes the molecules to move faster, and in order for the pressure not to increase, the volume of the gas must increase.

(c) Dalton's law. Dalton's law states the pressure of a mixture of gases is the sum of the pressures exerted by the individual gases. According to the kinetic molecular theory, there are no attractive forces between gas molecules; therefore, in a mixture of gases, each gas acts independently and the total pressure exerted will be the sum of the pressures exerted by the individual gases.

20.
$$N_2(g) + O_2(g) \longrightarrow 2\,NO(g)$$
$$1\ vol\ +\ 1\ vol\ \longrightarrow\ 2\ vol$$

According to Avogadro's Law, equal volumes of nitrogen and oxygen at the same temperature and pressure contain the same number of molecules. In the reaction, nitrogen and oxygen molecules react in a 1:1 ratio. Since two volumes of nitrogen monoxide are produced, one molecule of nitrogen and one molecule of oxygen must produce two molecules of nitrogen monoxide. Therefore, each nitrogen and oxygen molecule must be made up of two atoms.

21. We refer gases to STP because some reference point is needed to relate volume to moles. A temperature and pressure must be specified to determine the moles of gas in a given volume, and 0°C and 760 torr are convenient reference points.

22. Conversion of oxygen to ozone is an endothermic reaction. Evidence for this statement is that energy (286 kJ/3 mol O_2) is required to convert O_2 to O_3.

23. Oxygen atom = O   Oxygen molecule = O_2 Ozone molecule = O_3
An oxygen molecule contains 16 electrons.

24. Heating a mole of N_2 gas at constant pressure has the following effects:

(a) Density will decrease. Heating the gas at constant pressure will increase its volume. The mass does not change, so the increased volume results in a lower density.

(b) Mass does not change. Heating a substance does not change its mass.

(c) Average kinetic energy of the molecules increases. This is a basic assumption of the Kinetic Molecular Theory.

(d) Average velocity of the molecules will increase. Increasing the temperature increases the average kinetic energies of the molecules; hence, the average velocity of the molecules will increase also.

(e)   Number of N_2 molecules remains unchanged.   Heating does not alter the number of molecules present, except if extremely high temperatures were attained.   Then, the N_2 molecules might dissociate into N atoms resulting in fewer N_2 molecules.

25.   The "bends" are caused by a diver returning too rapidly to the surface.   The quick reduction in pressure causes the dissolved gases to form bubbles in the blood.

26.   Henry's law states that the amount of gas which will dissolve in a liquid is directly proportional to the pressure above the liquid.   This law is the basis for the "bends".   As the diver descends the pressure increases allowing more gas to dissolve in the blood.   As the diver ascends the pressure decreases the solubility of gas as well.   If the decrease is too rapid the "bends" result.

27.   The correct statements are b, d, f1, f4, h, i, n, o, p, q, t.

(a)   The pressure exerted by a gas, at constant volume, is dependent on the temperature of the gas.

(c)   At constant pressure, the volume of a gas is directly proportional to the absolute temperature.

(e)   Compressing a gas at constant temperature will cause its density to increase.   The mass will remain constant.

(j)   The volume of a gas depends upon its temperature, pressure, and amount of gas present.

(k)   In a mixture containing O_2 molecules and N_2 molecules, the O_2 molecules, on the average, are moving slower than the N_2 molecules.

(l) $PV = k$ is a statement of Boyle's law.

(m)   If the temperature of a sample of gas is increased from 25°C to 50°C, the volume of the gas will increase by about 8%.

(r)   According to the equation $2\ KClO_3(s) \xrightarrow{\Delta} 2\ KCl(s)\ +\ 3\ O_2(g)$

1 mol of $KClO_3$ will produce 33.6 L of O_2 at STP

(s) $PV = nRT$ is a statement of the ideal gas law.

28.   (a) $(715\text{ mm Hg})\left(\dfrac{1\text{ atm}}{760\text{ mm Hg}}\right)\ =\ 0.941\text{ atm}$

(b) $(715\text{ mm Hg})\left(\dfrac{1\text{ in. Hg}}{25.4\text{ mm Hg}}\right)\ =\ 28.1\text{ in. Hg}$

(c) $(715\text{ mm Hg})\left(\dfrac{14.7\text{ lb/in.}^2}{760\text{ mm Hg}}\right)\ =\ 13.8\text{ lb/in.}^2$

(d) $(715\text{ mm Hg})\left(\dfrac{1\text{ torr}}{1\text{ mm Hg}}\right)\ =\ 715\text{ torr}$

(e) $(715\text{ mm Hg})\left(\dfrac{1013\text{ mbar}}{760\text{ mm Hg}}\right)\ =\ 953\text{ mbar}$

(f) $(715\text{ mm Hg})\left(\dfrac{101.325\text{ kPa}}{760\text{ mm Hg}}\right)\ =\ 95.3\text{ kPa}$

29.   (a) $(28\text{ mm Hg})\left(\dfrac{1\text{ atm}}{760\text{ mm Hg}}\right)\ =\ 0.037\text{ atm}$

(b) $(6000.\text{ cm Hg})\left(\dfrac{1\text{ atm}}{76\text{ cm Hg}}\right)\ =\ 78.95\text{ atm}$

(c) $(795 \text{ torr})\left(\frac{1 \text{ atm}}{760 \text{ torr}}\right) = 1.05 \text{ atm}$

(d) $(5.00 \text{ kPa})\left(\frac{1 \text{ atm}}{101.325 \text{ kPa}}\right) = 0.0493 \text{ atm}$

30. In all cases $P_1V_1 = P_2V_2$ or $V_2 = \left(\frac{P_1V_1}{P_2}\right)$.

(a) $\frac{(500. \text{ mm Hg})(400. \text{ mL})}{760 \text{ mm Hg}} = 2.6 \times 10^2 \text{ mL}$

(b) $\frac{(500. \text{ torr})(400. \text{ mL})}{250 \text{ torr}} = 8.0 \times 10^2 \text{ mL}$

(c) $\frac{(500. \text{ mm Hg}) (1 \text{ atm}/760 \text{ mm Hg})(400. \text{ mL})}{2.00 \text{ atm}} = 132 \text{ mL}$

31. In all cases $P_1V_1 = P_2V_2$ or $V_2 = \left(\frac{P_1V_1}{P_2}\right)$.

(a) $\frac{(640. \text{ mm Hg})(500. \text{ mL})}{855 \text{ mL}} = 374 \text{ mm Hg}$

(b) $\frac{(640. \text{ mm Hg})(500. \text{ mL})}{450 \text{ mL}} = 7.1 \times 10^2 \text{ mm Hg}$

32. $P_1V_1 = P_2V_2$ or $P_2 = \left(\frac{P_1V_1}{V_2}\right)$.

$P_2 = \frac{(1.0 \text{ atm})(2500 \text{ L})}{25 \text{ L}} = 1.0 \times 10^2 \text{ atm}$

33. In each case $\frac{V_1}{T_1} = \frac{V_2}{T_2}$ or $V_2 = \frac{V_1T_2}{T_1}$;

Temperatures must be in Kelvin (°C + 273)

(a) $\frac{(6.00 \text{ L})(273 \text{ K})}{248 \text{ K}} = 6.60 \text{ L}$

(b) $\frac{(6.00 \text{ L})(255 \text{ K})}{248 \text{ K}} = 6.17 \text{ L}$

(c) $\frac{(6.00 \text{ L})(100. \text{ K})}{248 \text{ K}} = 2.42 \text{ L}$

(d) $\frac{(6.00 \text{ L})(345 \text{ K})}{248 \text{ K}} = 8.35 \text{ L}$

34. To double the volume of a gas, at constant pressure, the temperature (K) must be doubled.

$$\frac{V_1}{T_1} = \frac{V_2}{T_2} \qquad V_2 = 2\,V_1$$

$$\frac{V_1}{T_1} = \frac{2\,V_1}{T_2} \;\longrightarrow\; T_2 = \frac{2\,V_1 T_1}{V_1} \;\longrightarrow\; T_2 = 2\,T_1$$

$$T_2 = 2(300.\ K) = 600.\ K = 327°C$$

35. V = volume at 22°C and 740 torr

2 V = volume after change in temperature (P constant)

V = volume after change in pressure (T constant)

Since temperature is constant, $P_1 V_1 = P_2 V_2$ or $P_2 = \dfrac{P_1 V_1}{V_2}$

$$(740\ torr)\left(\frac{2\,V}{V}\right) = 1.5 \times 10^3\ torr \text{ (pressure to change 2 V to V)}$$

36. Volume is constant, so $\dfrac{P_1}{T_1} = \dfrac{P_2}{T_2}$ or $T_2 = \dfrac{T_1 P_2}{P_1}$;

$$T_2 = \frac{(500.\ torr)(295\ K)}{700.\ torr} = 211\ K = -62°C$$

37. The volume of the cylinder remains constant, so

$$\frac{P_1}{T_1} = \frac{P_2}{T_2} \text{ or } P_2 = \frac{P_1 T_2}{T_1} \text{ ;}$$

$$P_2 = \frac{(252\ atm)(77\ K)}{298\ K} = 65\ atm$$

38. The volume of the tires remain constant (until they burst), so

$$\frac{P_1}{T_1} = \frac{P_2}{T_2} \text{ or } T_2 = \frac{T_1 P_2}{P_1} \text{ ;}$$

$$T_2 = \frac{(44\ psi)(295\ K)}{30.\ psi} = 433\ K = 160°C = 320°F$$

39. Use the combined gas law $\dfrac{P_1 V_1}{T_1} = \dfrac{P_2 V_2}{T_2}$ or $V_2 = \dfrac{P_1 V_1 T_2}{P_2 T_1}$

(a) $\quad V_2 = \dfrac{(740\ mm\ Hg)(410\ mL)(273\ K)}{(760\ mm\ Hg)(300.\ K)} = 3.6 \times 10^2\ mL$

(b) $\quad V_2 = \dfrac{(740.\ mm\ Hg)(410.\ mL)(523\ K)}{(680\ mm\ Hg)(300.\ K)} = 7.8 \times 10^2\ mL$

40. Use the combined gas law $\dfrac{P_1V_1}{T_1} = \dfrac{P_2V_2}{T_2}$ or $V_2 = \dfrac{P_1V_1T_2}{P_2T_1}$

$$V_2 = \frac{(760 \text{ torr})(5.30 \text{ L})(343 \text{ K})}{(830 \text{ torr})(273 \text{ K})} = 6.1 \text{ L}$$

41. Use the combined gas law $\dfrac{P_1V_1}{T_1} = \dfrac{P_2V_2}{T_2}$ or $P_2 = \dfrac{P_1V_1T_2}{V_2T_1}$

$$P_2 = \frac{(1 \text{ atm})(800. \text{ mL})(303 \text{ K})}{(250. \text{ mL})(273 \text{ K})} = 3.55 \text{ atm}$$

42. Use the combined gas law $\dfrac{P_1V_1}{T_1} = \dfrac{P_2V_2}{T_2}$ or $V_2 = \dfrac{P_1V_1T_2}{P_2T_1}$

$$V_2 = \frac{(0.950 \text{ atm})(1400 \text{ mL})(275.0 \text{ K})}{(4.0 \text{ torr})(1 \text{ atm}/760 \text{ torr})(291 \text{ K})} = 2.4 \times 10^5 \text{ L}$$

43. Use the combined gas law $\dfrac{P_1V_1}{T_1} = \dfrac{P_2V_2}{T_2}$ or $V_2 = \dfrac{P_1V_1T_2}{P_2T_1}$

$$V_2 = \frac{(2.50 \text{ atm})(22.4 \text{ L})(268 \text{ K})}{(1.50 \text{ atm})(300. \text{ K})} = 33.4 \text{ L}$$

44. Use the combined gas law $\dfrac{P_1V_1}{T_1} = \dfrac{P_2V_2}{T_2}$ or $V_2 = \dfrac{P_1V_1T_2}{P_2T_1}$

First calculate the volume at STP.

$$V_2 = \frac{(400. \text{ torr})(600. \text{ mL})(273 \text{ K})}{(760. \text{ torr})(313 \text{ K})} = 275 \text{ mL}$$

At STP, a mole of any gas has a volume of 22.4 L

$$(0.275 \text{ L})\left(\frac{1 \text{ mol}}{22.4 \text{ L}}\right)\left(\frac{6.022 \times 10^{23} \text{ molecules}}{1 \text{ mol}}\right) = 7.39 \times 10^{21} \text{ molecules}$$

Each molecule contains 3 atoms, so:

$$(7.39 \times 10^{21} \text{ molecules})\left(\frac{3 \text{ atoms}}{1 \text{ molecule}}\right) = 2.22 \times 10^{22} \text{ atoms}$$

45. $P_{total} = P_{O_2} + P_{H_2O \text{ vapor}} = 720. \text{ torr}$

$P_{H_2O \text{ vapor}} = 17.5 \text{ torr}$

$P_{O_2} = 720. \text{ torr} - 17.5 \text{ torr} = 703 \text{ torr}$

46. $P_{total} = P_{N_2} + P_{H_2} + P_{O_2}$

$= 200. \text{ torr} + 600. \text{ torr} + 300. \text{ torr} = 1100 \text{ torr} = 1.10 \times 10^3 \text{ torr}$

47. $P_{total} = P_{CH_4} + P_{H_2O \text{ vapor}}$

$P_{H_2O \text{ vapor}} = 23.8 \text{ torr}$

$P_{CH_4} = 720. \text{ torr} - 23.8 \text{ torr} = 696 \text{ torr}$

To calculate the volume of dry methane, note the temperature is constant, so $P_1V_1 = P_2V_2$

$$V_2 = \frac{P_1V_1}{P_2} = \frac{(696 \text{ torr})(2.50 \text{ L})}{(760 \text{ torr})} = 2.29 \text{ L}$$

48. $CO_2 \qquad \dfrac{(5.00 \text{ L})(500. \text{ torr})}{10.0 \text{ L}} = 250. \text{ torr}$

$CH_4 \qquad \dfrac{(3.00 \text{ L})(400. \text{ torr})}{10.0 \text{ L}} = 120. \text{ torr}$

$P_{total} = P_{CO_2} + P_{CH_4} = 250. \text{ torr} + 120. \text{ torr} = 370. \text{ torr}$

49. $(2.5 \text{ mol})\left(\dfrac{22.4 \text{ L}}{\text{mol}}\right) = 56 \text{ L}$

50. The number of moles of a gas is proportional to pressure. (T and V constant)

(a) $(60. \text{ mol})\left(\dfrac{850 \text{ lb/in}^2}{1500 \text{ lb/in}^2}\right) = 34 \text{ mol}$

(b) $(60. \text{ mol})\left(\dfrac{2.016 \text{ g}}{\text{mol}}\right) = 1.2 \times 10^2 \text{ g } H_2$

51. $(2.5 \text{ L})\left(\dfrac{1 \text{ mol}}{22.4 \text{ L}}\right)\left(\dfrac{44.01 \text{ g}}{\text{mol}}\right) = 4.9 \text{ g } CO_2$

52. $(0.56 \text{ L})\left(\dfrac{1 \text{mol}}{22.4 \text{ L}}\right) = 0.025 \text{ mol}$

$\dfrac{1.08 \text{ g}}{0.025 \text{ mol}} = 43 \text{ g/mol (molar mass)}$

53. (a) $(1.0 \text{ mol})\left(\dfrac{22.4 \text{ L}}{\text{mol}}\right) = 22.4 \text{ L } NO_2$

(b) $(17.05 \text{ g})\left(\dfrac{1 \text{ mol}}{46.01 \text{ g}}\right)\left(\dfrac{22.4 \text{ L}}{\text{mol}}\right) = 8.30 \text{ L } NO_2$

(c) $(1.20 \times 10^{24} \text{ molecules})\left(\dfrac{1 \text{ mol}}{6.022 \times 10^{23} \text{ molecules}}\right)\left(\dfrac{22.4 \text{ L}}{\text{mol}}\right) = 44.6 \text{ L } NO_2$

54. $(1.00 \text{ L})\left(\dfrac{1 \text{ mol}}{22.4 \text{ L}}\right)\left(\dfrac{6.022 \times 10^{23} \text{ molecules}}{\text{mol}}\right) = 2.69 \times 10^{22} \text{ molecules}$

55. $(1.00 \text{ m}^3)\left(\dfrac{100 \text{ cm}}{1 \text{ m}}\right)^3\left(\dfrac{1 \text{ mL}}{1 \text{ cm}^3}\right)\left(\dfrac{1 \text{ L}}{1000 \text{ mL}}\right)\left(\dfrac{1 \text{ mol}}{22.4 \text{ L}}\right) = 44.6 \text{ mol } Cl_2$

56. $\left(\dfrac{1.78 \text{ g}}{\text{L}}\right)\left(\dfrac{22.4 \text{ L}}{\text{mol}}\right) = 39.9 \text{ g/mol (molar mass)}$

57. (a) $\quad d = \left(\dfrac{83.80 \text{ g}}{\text{mol}}\right)\left(\dfrac{1 \text{ mol}}{22.4 \text{ L}}\right) = 3.741 \text{ g/L}$

(b) $\quad d = \left(\dfrac{4.003 \text{ g}}{\text{mol}}\right)\left(\dfrac{1 \text{ mol}}{22.4 \text{ L}}\right) = 0.1787 \text{ g/L}$

(c) $\quad d = \left(\dfrac{80.06 \text{ g}}{\text{mol}}\right)\left(\dfrac{1 \text{ mol}}{22.4 \text{ L}}\right) = 3.574 \text{ g/L}$

(d) $\quad d = \left(\dfrac{56.10 \text{ g}}{\text{mol}}\right)\left(\dfrac{1 \text{ mol}}{22.4 \text{ L}}\right) = 2.504 \text{ g/L}$

58. (a) $\quad d = \left(\dfrac{38.00 \text{ g}}{\text{mol}}\right)\left(\dfrac{1 \text{ mol}}{22.4 \text{ L}}\right) = 1.696 \text{ g/L}$

(b) Assume 1 mol of F_2 and determine the volume using the ideal gas equation, $PV = nRT$.

$$V = \frac{nRT}{P} = \frac{(1.00 \text{ mol})(0.0821 \text{ L atm/mol K})(300. \text{ K})}{1.00 \text{ atm}} = 24.6 \text{ L}$$

$$D = \frac{38.00 \text{ g}}{24.6 \text{ L}} = 1.54 \text{ g/L}$$

59. At STP 22.4 L of CH_4 has a mass of 16.04 g. $\quad \dfrac{22.4 \text{ L}}{16.04 \text{ g}} = \dfrac{1.40 \text{ L}}{1.00 \text{ g}}$

Using 1.00 g as the mass of the sample:

$$P_1 = 1 \text{ atm} \qquad\qquad P_2 = 1 \text{ atm}$$
$$V_1 = 1.40 \text{ L} \qquad\qquad V_2 = 1.0 \text{ L}$$
$$T_1 = 273 \text{ K} \qquad\qquad T_2 = \text{?}$$

Since the pressure is constant, $\quad \dfrac{V_1}{T_1} = \dfrac{V_2}{T_2}$

$$T_2 = \frac{V_2 T_1}{V_1} = \frac{(1.00 \text{ L})(273 \text{ K})}{1.40 \text{ L}} = 195 \text{ K} = -78°C$$

60. (a) $\quad V = \dfrac{nRT}{P} = \dfrac{(0.510 \text{ mol})(0.0821 \text{ L atm/mol K})(320. \text{ K})}{1.6 \text{ atm}} = 8.4 \text{ L } H_2$

(b) $\quad n = \dfrac{PV}{RT} = \dfrac{(0.789 \text{ atm})(16.0 \text{ L})}{(0.0821 \text{ L atm/mol K})(300. \text{ K})} = 0.513 \text{ mol}$

The molar mass for CH_4 is 16.04 g/mol

$(16.04 \text{ g/mol})(0.513 \text{ mol}) = 8.23 \text{ g } CH_4$

(c) $PV = nRT$, but $n = \dfrac{m}{M}$ where M is the molar mass and m is the mass of the gas.

Thus, $PV = \dfrac{mRT}{M}$. To determine density, $d = m/V$.

Solving $PV = \dfrac{mRT}{M}$ for $\dfrac{m}{V}$ produces $\dfrac{m}{V} = \dfrac{PM}{RT}$.

$$d = \frac{m}{V} = \frac{(4.00 \text{ atm})(44 .01 \text{ g/mol})}{(0.0821 \text{ L atm/mol K})(253 \text{ K})} = 8.48 \text{ g/L}$$

(d)   Since $d = \frac{m}{V} = \frac{PM}{RT}$ from part (c), solve for M (molar mass)

$$M = \frac{DRT}{P} = \frac{(2.58 \text{ g/L})(0.0821 \text{ L atm/mol K})(300. \text{ K})}{1.00 \text{ atm}} = 63.5 \text{ g/mol (molar mass)}$$

61. $n = \frac{PV}{RT} = \frac{(0.813 \text{ atm})(0.215 \text{ L})}{(0.0821 \text{ L atm/mol K})(303 \text{ K})} = 7.03 \times 10^{-3} \text{ mol}$

molar mass $= \left(\frac{1.15 \text{ g}}{7.03 \times 10^{-3} \text{ mol}}\right) = 1.64 \times 10^2 \text{ g/mol}$

62. $V = \frac{nRT}{P} = \frac{(2.3 \text{ mol})(0.0821 \text{ L atm/mol K})(300. \text{ K})}{(750/760) \text{ atm}} = 57 \text{ L Ne}$

63. When working with gases, the identity of the gas does not affect the volume, as long as the number of moles are known.

Total moles $= \text{mol}_{H_2} + \text{mol}_{CO_2} = 5.00 \text{ mol} + 0.500 \text{ mol} = 5.50 \text{ mol}$

$V = (5.50 \text{ mol})\left(\frac{22.4 \text{ L}}{\text{mol}}\right) = 123 \text{ L}$

64. $T = \frac{PV}{nR} = \frac{(4.15 \text{ atm})(0.250 \text{ L})}{(4.50 \text{ mol})(0.0821 \text{ L atm/mol K})} = 2.81 \text{ K}$

65. $n = \frac{PV}{RT} = \frac{(0.500 \text{ atm})(5.20 \text{ L})}{(0.0821 \text{ L atm/mol K})(250 \text{ K})} = 0.13 \text{ mol N}_2$

66. $(8.30 \text{ mol Al})\left(\frac{3 \text{ mol H}_2}{2 \text{ mol Al}}\right)\left(\frac{22.4 \text{ L}}{\text{mol}}\right) = 279 \text{ L H}_2 \text{ at STP}$

67. (a) $(5.5 \text{ mol NO})\left(\frac{4 \text{ mol NH}_3}{4 \text{ mol NO}}\right) = 5.5 \text{ mol NH}_3$

(b) $(7.0 \text{ mol O}_2)\left(\frac{4 \text{ mol NH}_3}{5 \text{ mol O}_2}\right) = 5.6 \text{ mol NH}_3$

(c)   Limiting reactant problem.   Remember, volume - volume relationships are the same as mole - mole relationships when dealing with gases.

$(12 \text{ L O}_2)\left(\frac{4 \text{ mol NO}}{5 \text{ mol O}_2}\right) = 9.6 \text{ L NO (from O}_2)$

$(10. \text{ L NH}_3)\left(\frac{4 \text{ mol NO}}{4 \text{ mol NH}_3}\right) = 10. \text{ L NO (from NH}_3)$

Oxygen is the limiting reactant, 9.6 L NO is formed.

(d) $(800. \text{ mL O}_2)\left(\frac{4 \text{ mol NO}}{5 \text{ mol O}_2}\right) = 640. \text{ mL NO} = 0.640 \text{ L NO}$

(e)   Limiting reactant problem.   Remember, volume - volume relationships are the same as mole - mole relationships when dealing with gases.

$$(3.0 \text{ L NH}_3)\left(\frac{4 \text{ mol NO}}{4 \text{ mol NH}_3}\right) = 3.0 \text{ L NO (from NH}_3)$$

$$(3.0 \text{ L O}_2)\left(\frac{4 \text{ mol NO}}{5 \text{ mol O}_2}\right) = 2.4 \text{ L NO (from O}_2)$$

Oxygen is the limiting reactant, 2.4 L NO is formed.

(f) $$(60. \text{ L NO})\left(\frac{1 \text{ mol}}{22.4 \text{ L}}\right)\left(\frac{5 \text{ mol O}_2}{4 \text{ mol NO}}\right)\left(\frac{32.00 \text{ g}}{\text{mol}}\right) = 1.1 \times 10^2 \text{ g O}_2$$

(g) $4 \text{ L NH}_3 \longrightarrow 4 \text{ L NO} + 6 \text{ L H}_2\text{O}$ (10 L product)

$$(32 \text{ L product})\left(\frac{4 \text{ L NH}_3}{10 \text{ L product}}\right)\left(\frac{1 \text{ mol}}{22.4 \text{ L}}\right)\left(\frac{17.03 \text{ g}}{\text{mol}}\right) = 9.7 \text{ g NH}_3$$

68. The balanced equation is

$$4 \text{ FeS} + 7 \text{ O}_2 \xrightarrow{\Delta} 2 \text{ Fe}_2\text{O}_3 + 4 \text{ SO}_2$$

(a) $$(600. \text{ g FeS})\left(\frac{1 \text{ mol}}{87.91 \text{ g}}\right)\left(\frac{7 \text{ mol O}_2}{4 \text{ mol FeS}}\right)\left(\frac{22.4 \text{ L}}{\text{mol}}\right) = 268 \text{ L O}_2$$

(b) $$(600. \text{ g FeS})\left(\frac{1 \text{ mol}}{87.91 \text{ g}}\right)\left(\frac{4 \text{ mol SO}_2}{4 \text{ mol FeS}}\right)\left(\frac{22.4 \text{ L}}{\text{mol}}\right) = 153 \text{ L SO}_2$$

69. $1.0 \text{ mol C}_2\text{H}_2 \longrightarrow 1.0 \text{ mol C}_2\text{H}_4\text{F}_2$

$$(5.0 \text{ mol HF})\left(\frac{1 \text{ mol C}_2\text{H}_4\text{F}_2}{2 \text{ mol HF}}\right) = 2.5 \text{ mol C}_2\text{H}_4\text{F}_2$$

C_2H_2 is the limiting reactant. 1.0 mol $C_2H_4F_2$ forms, no moles C_2H_2 remain. According to the equation, 2.0 mol HF yields 1.0 mol $C_2H_4F_2$. Therefore,

5.0 mol HF − 2.0 mol HF = 3.0 mol HF unreacted

The flask contains 1.0 mole $C_2H_4F_2$ and 3.0 mol HF when the reaction is complete.

$$P = \frac{nRT}{V} = \frac{(4.0 \text{ mol})(0.0821 \text{ L atm/mol K})(273 \text{ K})}{10.0 \text{ L}} = 9.0 \text{ atm}$$

70. According to Graham's Law of Effusion, the rates of effusion are inversely proportional to the molar mass.

$$\frac{\text{rate He}}{\text{rate N}_2} = \sqrt{\frac{\text{molar mass N}_2}{\text{molar mass He}}} = \sqrt{\frac{28.02}{4.003}} = \sqrt{7.0} = 2.646$$

Helium effuses 2.646 times faster than nitrogen.

71. (a) $$\frac{\text{rate CH}_4}{\text{rate He}} = \sqrt{\frac{16.04}{4.003}} = \sqrt{4.0} = 2.0$$

Helium effuses twice as fast as CH_4.

(b) x = distance He travels

$100 - x$ = distance CH_4 travels

D_{He} = $2\,D_{CH_4}$ D = distance traveled

x = $2(100 - x)$

$3x$ = 200

x = 66.7 cm

The gases meet 66.7 cm from the helium end.

72. C $(85.7\ g)\left(\dfrac{1\ mol}{12.01\ g}\right)$ = 7.14 mol $\dfrac{7.14}{7.14}$ = 1 mol

H $(14.3\ g)\left(\dfrac{1\ mol}{1.008\ g}\right)$ = 14.19 mol $\dfrac{14.19}{7.14}$ = 1.99 mol

The empirical formula is CH_2. To determine the molecular formula, the molar mass must be known.

$\left(\dfrac{2.50\ g}{L}\right)\left(\dfrac{22.4\ L}{mol}\right)$ = 56.0 g/mol

The empirical formula mass is 14.0. $\dfrac{56.0}{14.0}$ = 4

Therefore, the molecular formula is $(CH_2)_4$ = C_4H_8

73. $(10.0\ mol\ CO)\left(\dfrac{2\ mol\ CO_2}{2\ mol\ CO}\right)$ = 10.0 mol CO_2 (from CO)

$(8.0\ mol\ O_2)\left(\dfrac{2\ mol\ CO_2}{1\ mol\ O_2}\right)$ = 16 mol CO_2 (from O_2)

CO: the limiting reactant, O_2: in excess, 3.0 mole O_2 unreacted

(a) 10.0 mol CO_2 present, 3.0 mol O_2, and no CO will be present.

(b) P = $\dfrac{nRT}{V}$ = $\dfrac{(13\ mol)(0.0821\ L\ atm/mol\ K)(273\ K)}{10.\ L}$ = 29 atm

74. $(0.25\ L\ O_2)\left(\dfrac{1\ mol}{22.4\ L}\right)$ = 5.6 mol O_2

$(5.6\ mol\ O_2)\left(\dfrac{2\ mol\ KClO_3}{3\ mol\ O_2}\right)\left(\dfrac{122.6\ g}{mol}\right)$ = 0.91 g $KClO_3$ in the sample

$\left(\dfrac{0.91\ g}{1.20\ g}\right)(100)$ = 76% $KClO_3$

75. Some ammonia gas dissolves in the water squirted into the flask, lowering the pressure inside the flask. The atmospheric pressure is greater than the pressure inside the flask and pushes water from the beaker up the tube and into the flask, filling the flask.

76. (a) The pressure of the helium is simply the difference between levels of Hg, or 250 mm Hg (250 torr).

(b) The pressure of the oxygen is the difference between the pressure of the atmosphere and the difference in the levels of Hg.

$$P_{O_2} = P_{atm} + 300 \text{ mm Hg}$$
$$= 760 \text{ mm Hg} + 300 \text{ mm Hg}$$
$$= 1060 \text{ mm Hg (1060 torr)}$$

77. Each gas behaves as though it were alone in a 4.0 L system.

(a) After expansion: $P_1V_1 = P_2V_2$

For CO_2 $\quad P_2 = \dfrac{P_1V_1}{V_2} = \dfrac{(150. \text{ torr})(3.0 \text{ L})}{4.0 \text{ L}} = 1.1 \times 10^2 \text{ torr}$

For H_2 $\quad P_2 = \dfrac{P_1V_1}{V_2'} = \dfrac{(50. \text{ torr})(1.0 \text{ L})}{4.0 \text{ L}} = 13 \text{ torr}$

$P_{total} = P_{H_2} + P_{CO_2} = 110 \text{ torr} + 13 \text{ torr} = 120 \text{ torr}$

78. Assume 1.00 L of air. The mass of the sample is 1.29 g.

$$\frac{P_1V_1}{T_1} = \frac{P_2V_2}{T_2}$$

$$V_2 = \frac{P_1V_1T_2}{P_2T_1} = \frac{(760 \text{ torr})(1.00 \text{ L})(290. \text{ K})}{(450 \text{ torr})(273 \text{ K})} = 1.8 \text{ L}$$

$$d = \frac{m}{V} = \frac{1.29 \text{ g}}{1.8 \text{ L}} = 0.72 \text{ g/L}$$

79. Air enters the room. The volume is directly proportional to temperature.

Initial volume of the room $= V_1 = (16 \text{ ft})(12 \text{ ft})(12 \text{ ft})$
$$= 2300 \text{ ft}^3$$

Final volume of the room (after change in temperature)

$$V_2 = \frac{V_1T_2}{T_1} = \frac{(2300 \text{ ft}^3)(270 \text{ K})}{300 \text{ K}} = 2100 \text{ ft}^3$$

Since the volume of the gas is less at the new temperature, air would enter the room.

80. $\quad P_1 = 40.0 \text{ atm} \qquad P_2 = ?$

$V_1 = 50.0 \text{ L} \qquad V_2 = 50.0 \text{ L}$

$T_1 = 25 \text{ °C} = 298 \text{ K} \qquad T_2 = 25°\text{C} + 152°\text{C} = 177°\text{C} = 450. \text{ K}$

Gas cylinders have constant volume, so pressure and temperature vary directly.

$$P_2 = \frac{P_1T_2}{T_1} = \frac{(40.0 \text{ atm})(450. \text{ K})}{298 \text{ K}} = 60.4 \text{ atm}$$

CHAPTER 14

WATER AND THE PROPERTIES OF LIQUIDS

1. The potential energy is greater in the liquid water than in the ice. The heat necessary to melt the ice increases the potential energy, thus allowing the molecules greater freedom of motion.

2. At 0°C, all three substances, H_2S, H_2Se, and H_2Te, are gases, because they all have boiling points below 0°C.

3. The pressure of the atmosphere must be 1.00 atmosphere, otherwise the water would be boiling at some other temperature.

4.

5. Melting point, boiling point, heat of fusion, heat of vapoization, density, and crystallization structure in the solid state are some of the physical properties of water that would be very different, if the the molecules were linear and nonpolar instead of bent and highly polar.

6. Prefixes preceding the word hydrate are used in naming hydrates, indicating the number of molecules of water present in the formulas. The prefixes used are:

    mono = 1   di = 2   tri = 3   tetra = 4   penta = 5
    hexa = 6   hepta = 7   octa = 8   nona = 9   deca = 10

7. The distillation setup in Figure 14.10 would be satisfactory for separating salt and water, but not for separating ethyl alcohol and water. In the first case, the water is easily vaporized, the salt is not, so the water boils off and condenses. In the second case, both substances are easily vaporized, so both would vaporize (though not to an equal degree) and the condensed liquid would contain both substances.

8. The ethyl alcohol would be at 70°C. The liquid is boiling, which means its vapor pressure equals the confining pressure. From Table 14.1, we find that ethyl alcohol has a vapor pressure of 543 torr at 70°C.

9. The water in the containers would both have the same vapor pressure, for it is a function of the temperature of the liquid. The pressure exerted by water vapor molecules would be less in the open container, because the opportunity to escape prevents the establishment of equilibrium.

10. In Figure 14.1, it would be case (b) in which the atmosphere would reach saturation.

11. If ethyl ether and ethyl alcohol were both placed in a closed container, (a) both substances would be present in the vapor, for both are volatile liquids: (b) ethyl ether would have more molecules in the vapor because it has a higher vapor pressure.

12. The equilibrium vapor pressure observed in (c) would remain unchanged. The presence of more water in (b) does not change the magnitude of the equilibrium vapor pressure of the water. The temperature of the water determines the magnitude of the equilibrium vapor pressure.

13. At 30 torr, H_2O would boil at approximately 29°C, ethyl alcohol at 14°C, and ethyl ether and ethyl chloride at some temperature below 0°C.

14. (a) At a pressure of 500 torr, water boils at 88°C.

    (b) The normal boiling point of ethyl alcohol is 78°C.

    (c) At a pressure of 0.50 atm (380 torr), ethyl ether boils at 16°C.

15. Based on Figure 14.5:

    (a) Line BC is horizontal because the temperature remains constant during the entire process of melting.

    (b) During BC, both solid and liquid phases will be present.

    (c) The line DE represents the change from liquid to vapor, that is, vaporization. The horizontal line is at the boiling temperature of the substance.

16. Physical properties of water:

    (a)   melting point, 0°C
    (b)   boiling point, 100°C (at 1 atm pressure)
    (c)   colorless
    (d)   odorless
    (e)   tasteless
    (f)   heat of fusion, 335 J/g (80 cal/g)
    (g)   heat of vaporization, 2.26 kJ/g (540 cal/g)
    (h)   density = 1.0 g/mL (at 4°C)
    (i)   specific heat = 4.184 J/g°C

17. For water, to have its maximum density, the temperature must be 4°C, and the pressure sufficient to keep it liquid. $d = 1.0$ g/mL

18. If you apply heat to an ice-water mixture, the heat is transferred to melt ice, rather than warm the water, so the temperature remains constant until all the ice has melted.

19. Ice at 0°C contains less heat than water at 0°C. Heat must be added to convert ice to water, so the water will contain that much additional heat.

20. Ice floats in water because it is less dense than water. The density of ice at 0°C is 0.915 g/mL. Liquid water, however, has a density of 1.00 g/mL. Ice will sink in ethyl alcohol, which has a density of 0.79 g/mL.

21. The heat of vaporization of water would be lower if water molecules were linear instead of bent. If linear, the molecules of water would be nonpolar. The relatively high heat of vaporization of water is a result of the molecule being highly polar and having strong dipole-dipole attraction for other water molecules.

22. Ethyl alcohol exhibits hydrogen bonding; ethyl ether does not. This is indicated by the high heat of vaporization of ethyl alcohol, even though its molar mass is much less than the molar mass of ethyl ether.

23. A linear water molecule, being nonpolar, would exhibit less hydrogen bonding than the highly polar, bent, water molecule. The polar molecule has a greater attractive force than a nonpolar molecule.

24. Ammonia exhibits hydrogen bonding; methane does not. The ammonia molecule is polar; the methane molecule is not. The nitrogen atom in ammonia is quite electronegative; the carbon atom in methane is much less electronegative.

25. Water, at 80°C, will have fewer hydrogen bonds than water at 40°C. At the higher temperature, the molecules of water are moving faster than at the lower temperature. This results in less hydrogen bonding at the higher temperature.

26. $H_2NCH_2CH_2NH_2$ has two polar NH_2 groups. It should, therefore, show more hydrogen bonding and a higher boiling point (49°C).

27. Rubbing alcohol feels cold when applied to the skin, because the evaporation of the alcohol absorbs heat from the skin. The alcohol has a fairly high vapor pressure and evaporates quite rapidly. This produces the cooling effect.

28. (a)   Order of increasing rate of evaporation: Mercury, acetic acid, water, toluene, benzene, carbon tetrachloride, methyl alcohol, bromine.

    (b)   Highest boiling point is mercury. Lowest boiling point is bromine.

29. Water boils when its vapor pressure equals the prevailing atmospheric pressure over the water. In order for water to boil at 50°C, the pressure over the water would need to be reduced to a point equal to the vapor pressure of the water (92.5 torr).

30. Two conditions must be met for water molecules to evaporate:

    (1)   The evaporating molecules must be on the surface of the water.

    (2)   The molecules must have sufficient kinetic energy to overcome attractive forces of other water molecules. Consequently, higher energy molecules escape; lower energy molecules are left behind. Evaporation, thus cools the remaining liquid.

31. In a pressure cooker, the temperature at which the water boils increases above its normal boiling point, because the water vapor (steam) formed by boiling cannot escape. This results in an increased pressure over water and, consequently, an increased boiling temperature.

32. Vapor pressure varies with temperature. The temperature at which the vapor pressure of a liquid equals the prevailing pressure is the boiling point.

33. As temperature increases, molecular velocities increase. At higher velocities, it becomes easier for molecules to break away from the attractive forces in the liquid.

34. Water has a relatively high boiling point because there is a high attraction between molecules due to hydrogen bonding.

35. Ammonia would have a higher vapor pressure than SO_2 at –40°C.

36. As the temperature of a liquid increases, the kinetic energy of the molecules as well as the vapor pressure of the liquid increases. When the vapor pressure of the liquid equals the external pressure, boiling begins with many of the molecules having enough energy to escape from the liquid. Bubbles of vapor are formed throughout the liquid and these bubbles rise to the surface, escaping as boiling continues.

37. HF has a higher boiling point that HCl because of the strong hydrogen bonding in HF. Neither F_2 nor Cl_2 will have hydrogen bonding, so the compound, F_2, with the smaller mass, has the lower boiling point.

38. Yes. Ice can be cooled below 0°C, the same as any other solid.

39. The boiling liquid remains at constant temperature because the added heat energy is being used to convert the liquid to a gas, i.e., to supply the heat of vaporization for the liquid at its boiling point.

40. 2336°C, the boiling point of copper.

41. The lake freezes from the top down because, as the temperature drops to freezing or below, the water on the surface tends to cool faster than the water that lies deeper. As the surface water freezes, the ice formed floats because the ice is less dense than the liquid water below it.

42. If the lake is in an area where the temperature is below freezing for part of the year, the expected temperature would be 4°C at the bottom of the lake. This is because the surface water would cool to 4°C (maximum density) and sink.

43. Reactions of metals with water:

$2 Al + 3 H_2O$ (steam) $\longrightarrow$ $3 H_2(g) + Al_2O_3$ requires steam

$Ca + 2 H_2O \longrightarrow H_2(g) + Ca(OH)_2$ slowly at room temperature

$3 Fe + 4 H_2O$ (steam) $\longrightarrow$ $4 H_2(g) + Fe_3O_4$ requires steam

$2 Na + 2 H_2O \longrightarrow H_2(g) + 2 NaOH$ vigorously at room temp

$Zn + H_2O$ (steam) $\longrightarrow$ $H_2(g) + ZnO$ requires steam

44. The formation of hydrogen and oxygen from water is an endothermic reaction, due to the following evidence:

(a) Energy must be continually be provided to the system for the reaction to proceed. The reaction will cease when the energy source is removed.

(b) The reverse reaction, burning hydrogen in oxygen, releases energy as heat.

45. (a) The word anhydride originates from the Greek, anhydros, meaning waterless. An anhydride is a substance that will react with water to form an acid or base.

(b) An acid anhydride will be an oxide of a nonmetal.

(c) A basic anhydride will be an oxide of a metal.

46.   (a)   Acid anyhdride: $[H_2SO_3 \ SO_2]$ $[H_2SO_4 \ SO_3]$ $[HNO_3 \ N_2O_5]$
                $[HClO_4 \ Cl_2O_7]$ $[H_2CO_3 \ CO_2]$ $[H_3PO_4 \ P_2O_5]$

     (b)   Basic anhydrides: $[NaOH \ Na_2O]$ $[KOH \ K_2O]$ $[Ba(OH)_2 \ BaO]$
                $[Ca(OH)_2 \ CaO]$ $[Mg(OH)_2 \ MgO]$

47.   (a) $Ba(OH)_2 \xrightarrow{\Delta} BaO + H_2O$

     (b) $2\,CH_3OH + 3\,O_2 \longrightarrow 2\,CO_2 + 4\,H_2O$

     (c) $2\,Rb + 2\,H_2O \longrightarrow 2\,RbOH + H_2$

     (d) $SnCl_2 \cdot 2\,H_2O \xrightarrow{\Delta} SnCl_2 + 2\,H_2O$

     (e) $HNO_3 + NaOH \longrightarrow NaNO_3 + H_2O$

     (f) $Li_2O + H_2O \longrightarrow 2\,LiOH$

     (g) $2\,KOH \xrightarrow{\Delta} K_2O + H_2O$

     (h) $Ba + 2\,H_2O \longrightarrow Ba(OH)_2 + H_2$

     (i) $Cl_2 + H_2O \longrightarrow HCl + HClO$

     (j) $SO_3 + H_2O \longrightarrow H_2SO_4$

     (k) $H_2SO_3 + 2\,KOH \longrightarrow K_2SO_3 + 2\,H_2O$

     (l) $CO_2 + H_2O \longrightarrow H_2CO_3$

48.   (a)   barium bromide dihydrate
     (b)   aluminum chloride hexahydrate
     (c)   iron(III) phosphate tetrahydrate
     (d)   magnesium ammonium phosphate hexahydrate
     (e)   iron(II) sulfate heptahydrate
     (f)   tin(IV) chloride pentahydrate

49. $CuSO_4$ (anhydrous) is gray white. When exposed to moisture, it turns bright blue ($CuSO_4 \cdot 5\,H_2O$). The color change is an indicator of moisture in the environment.

50. $MgSO_4 \cdot 7\,H_2O$ $Na_2HPO_4 \cdot 12\,H_2O$

51. Deionized water is water from which the ions have been removed.

     (a)   Hard water contains dissolved calcium and magnesium salts.

     (b)   Soft water is free of ions that cause hardness.

     (c)   Distilled water has been vaporized by boiling and recondensed. It is free of nonvolatile impurities, but may still contain any volatile impuriities that were initially present in the water.

52. Soap can soften hard water by forming a precipitate with, and thus removing, the calcium and magnesium ions. This precipitate is a greasy scum and is very objectional, so it is a poor way to soften water.

53. Chlorine is commonly used to destroy bacteria in water. Ozone is also used in some places.

54. Ozone

55. When organic pollutants in water are oxidized by dissolved oxygen, there may not be sufficient dissolved oxygen to sustain marine life, such as fish. Most marine life forms depend on dissolved oxygen for cellular respiration.

56. Liquids that are stored in ceramic containers should never be drunk, for they are likely to have dissolved some of the lead from the ceramic. If the ceramic is glazed, the liquid is less apt to dissolve substances from the ceramic.

57. Na_2 zeolite*(s)* $+$ Mg^{2+}*(aq)* $\longrightarrow$ Mg zeolite*(s)* $+$ $2\,Na^+$*(aq)*

58. Softening of hard water using sodium carbonate:

$$CaCl_2\text{\textit{(aq)}} + Na_2CO_3\text{\textit{(aq)}} \longrightarrow CaCO_3\text{\textit{(s)}} + 2\,NaCl\text{\textit{(aq)}}$$

59. The correct statements are a, b, c, f, h, l, m, o, p, s, t, u, w, y.

   (d)   The changing of ice into water is an endothermic process.

   (e)   Water and hydrogen fluoride are both polar molecules.

   (g) $2\,H_2O_2 \longrightarrow 2\,H_2O + O_2$ represents a balanced equation for the decomposition of hydrogen peroxide.

   (i)   The density of water is dependent on temperature.

   (j)   Liquid A boils at a lower temperature than liquid B. This fact indicates that liquid A has a higher vapor pressure than liquid B at any particular temperature.

   (k)   Water boils at a lower temperature in the mountains than at sea level.

   (n)   The normal boiling temperature of water is 100°C.

   (q)   Calcium oxide reacts with water to form calcium hydroxide.

   (r)   Carbon dioxide is the anhydride of carbonic acid.

   (v)   Distillation is effective for softening water because the pure water is distilled, leaving the minerals behind.

   (x)   Disposal of toxic industrial wastes in toxic waste dumps has been found to be an unsatisfactory long-term solution to the problem.

   (z) $CaCl_2 \cdot 2\,H_2O$ has a higher percentage of water than $BaCl_2 \cdot 2\,H_2O$.

60. Humectants and emollients soften the skin by increasing its water content. Humectants attract water vapor from the air, while emollients cover the skin with a layer of material which is immiscible with water.

61. The triangle theory of sweetness states that molecules which are sweet contain 3 specific sites which produce the proper structure to attach to the taste buds and trigger the "sweet" response.

62. A good sweetener is:

    1.   as sweet or sweeter than sugar (sucrose)
    2.   nontoxic
    3.   quick to register sweet on the taste buds
    4.   easy to release
    5.   noncaloric
    6.   stable when cooked or dissolved
    7.   inexpensive

63. Humectants are polar compounds while emollients are nonpolar compounds.

64. (a) $(100. \text{ g CoCl}_2 \cdot 6 \text{ H}_2\text{O})\left(\frac{1 \text{ mol}}{238.0 \text{ g}}\right) = 0.420 \text{ mol CoCl}_2 \cdot 6 \text{ H}_2\text{O}$

      (b) $(100. \text{ g FeI}_2 \cdot 4 \text{ H}_2\text{O})\left(\frac{1 \text{ mol}}{381.7 \text{ g}}\right) = 0.262 \text{ mol FeI}_2 \cdot 4 \text{ H}_2\text{O}$

65. (a) $(100. \text{ g CoCl}_2 \cdot 6 \text{ H}_2\text{O})\left(\frac{1 \text{ mol}}{238.0 \text{ g}}\right)\left(\frac{6 \text{ mol H}_2\text{O}}{1 \text{ mol CoCl}_2 \cdot 6 \text{ H}_2\text{O}}\right) = 2.52 \text{ mol H}_2\text{O}$

      (b) $(100. \text{ g FeI}_2 \cdot 4 \text{ H}_2\text{O})\left(\frac{1 \text{ mol}}{381.7 \text{ g}}\right)\left(\frac{4 \text{ mol H}_2\text{O}}{1 \text{ mol FeI}_2 \cdot 4 \text{ H}_2\text{O}}\right) = 1.05 \text{ mol H}_2\text{O}$

66. Assume 1 mol of the compound.

$$\left(\frac{\text{g H}_2\text{O}}{\text{g MgSO}_4 \cdot 7 \text{ H}_2\text{O}}\right)(100) = \left(\frac{(7)(18.02 \text{ g})}{246.4 \text{ g}}\right)(100) = 51.2\% \text{ H}_2\text{O}$$

67. Assume 1 mol of hydrate.

$$\% \text{ H}_2\text{O} = \frac{\text{g H}_2\text{O}}{\text{g Al}_2(\text{SO}_4)_3 \cdot 18 \text{ H}_2\text{O}} = \left(\frac{(18)(18.02 \text{ g})}{666.5 \text{ g}}\right)(100) = 48.7\% \text{ H}_2\text{O}$$

68. Assume 100. g of the compound.

    $(0.142)(100. \text{ g}) = 14.2 \text{ g H}_2\text{O}$

    $(0.858)(100. \text{ g}) = 85.8 \text{ g Pb(C}_2\text{H}_3\text{O}_2)_2$

    $(14.2 \text{ g H}_2\text{O})\left(\frac{1 \text{ mol}}{18.02 \text{ g}}\right) = 0.789 \text{ mol H}_2\text{O}$

    $(85.8 \text{ g Pb(C}_2\text{H}_3\text{O}_2)_2)\left(\frac{1 \text{ mol}}{325.3 \text{ g}}\right) = 0.264 \text{ mol Pb(C}_2\text{H}_3\text{O}_2)_2$

In the formula for the hydrate, there is one mole of $\text{Pb(C}_2\text{H}_3\text{O}_2)_2$, so divide each of the moles by 0.264.

$$\frac{0.264 \text{ mol Pb(C}_2\text{H}_3\text{O}_2)_2}{0.264 \text{ mol}} = 1 \text{ mol Pb(C}_2\text{H}_3\text{O}_2)_2$$

$$\frac{0.789 \text{ mol H}_2\text{O}}{0.264 \text{ mol}} = 2.99 \text{ mol H}_2\text{O}$$

Therefore, the formula is $\text{Pb(C}_2\text{H}_3\text{O}_2)_2 \cdot 3 \text{ H}_2\text{O}$.

69.   25.0 g hydrate $-$ 16.9 g $FePO_4$ = 8.1 g H_2O

$(8.1 \text{ g } H_2O)\left(\frac{1 \text{ mol}}{18.02 \text{ g}}\right)$ = 0.45 mol H_2O $\frac{0.45}{0.112}$ = 4

$(16.9 \text{ g } FePO_4)\left(\frac{1 \text{ mol}}{150.8 \text{ g}}\right)$ = 0.112 mol $FePO_4$ $\frac{0.112}{0.112}$ = 1

The formula is $FePO_4 \cdot 4\ H_2O$

70. (a)   Warm water   20.°C $\longrightarrow$ 100.°C

E_a = (m)(specific heat)(ΔT) = $(120. \text{ g})\left(\frac{4.184 \text{ J}}{\text{g }°\text{C}}\right)(80.°\text{C}) = 4.0 \times 10^4$ J

(b)   Convert water to steam

E_b = (m)(heat of vaporization) = $(120. \text{ g})(2.26 \times 10^3 \text{ J/g}) = 2.71 \times 10^5$ J

E_{total} = E_a + E_b = $(4.0 \times 10^4 \text{ J})$ + $(2.71 \times 10^5 \text{ J}) = 3.11 \times 10^5$ J

71. (a)   Cool water   24°C $\longrightarrow$ 0°C

E_a = (m)(specific heat)(ΔT) = $(126 \text{ g})\left(\frac{4.184 \text{ J}}{\text{g }°\text{C}}\right)(24°\text{C}) = 1.3 \times 10^4$ J

(b)   Convert water to ice

E_b = (m)(heat of fusion) = $(126 \text{ g})(335 \text{ J/g})$ = 4.22×10^4 J

E_{total} = E_a + E_b = $(1.3 \times 10^4 \text{ J})$ + $(4.22 \times 10^4 \text{ J}) = 5.5 \times 10^4$ J

72. (a)   Melt ice: E_a = (m)(heat of fusion) = $(225 \text{ g})(80. \text{ cal/g})$ = 18,000 cal

(b)   Warm the water: E_b = (m)(specific heat)(ΔT) = $(225 \text{ g})\left(\frac{1 \text{ cal}}{\text{g }°\text{C}}\right)(100.°\text{C})$ = 22,500 cal

(c)   Vaporize water: E_c = (m)(heat of vaporization) = $(225 \text{ g})(540 \text{ cal/g})$ = 121,500 cal

E_{total} = E_a + E_b + E_c = 1.62×10^5 cal

73. $(2.26 \text{ kJ/g})\left(\frac{18.02 \text{ g}}{\text{mol}}\right)$ = 40.7 kJ/mol

74.   Heat released in cooling water

E = (m)(specific heat)(ΔT) = $(300. \text{ g})\left(\frac{1 \text{ cal}}{\text{g }°\text{C}}\right)(25°\text{C})$ = 7500 cal

Heat required to melt ice

E = (m)(heat of fusion) = $(100 \text{ g})(80 \text{ cal/g})$ = 8000 cal

Less energy is released in cooling the water than is required to melt the ice.   Ice will remain and the water will be at 0°C.

75.   Heat lost by warm water = heat gained by the ice

x = final temperature

$(1500 \text{ g})\left(\frac{1 \text{ cal}}{\text{g }°\text{C}}\right)(75°\text{C} - x)$ = $(75 \text{ g})(80 \frac{\text{cal}}{\text{g}})$ + $(75 \text{ g})\left(\frac{1 \text{ cal}}{\text{g }°\text{C}}\right)(x - 0°\text{C})$

(112,500 cal) $-$ (1500x cal/°C) = 6000 cal + 75x cal/°C

106,500 cal = 1575x cal/°C

68°C = x

76. Assume that room temperature is 20.°C. Then

$E = (m)(\text{specific heat})(\Delta T) = (250.\ g)\left(\frac{0.096\ cal}{g\ °C}\right)(150.\ -\ 20.°C) = 3.1 \times 10^3\ cal$ (3.1 kcal)

77. $E = (m)(\text{heat of fusion})$

$9560\ J = (500.\ g)(335\ J/g)$

$9560\ J < 167,500\ J$

Since 167,500 J are required to melt all the ice, and only 9560 J are available, the system will be at 0°C. It will be a mixture of ice and water.

78. Heat lost by warm water = heat gained by ice

$$(120.\ g)\left(\frac{1\ cal}{g\ °C}\right)(45°C - 0°C) = (m)(80\ cal/g)$$

$68\ g = m$ (grams of ice melted)

68 g of ice melted. Therefore, 150. g − 68 g = 82 g ice remains.

79. Energy to heat water = energy to condense steam

$(300.\ g)\left(\frac{1\ cal}{g\ °C}\right)(100°C\ -\ 25°C) = (m)(540\ cal/g)$

$42\ g = m$ (grams of steam to heat the water to 100°C)

42 g of steam are required to heat the water to 100°C. Since only 35 g of steam are added to the system, the final temperature will be less than 100°C.

80. $(50.0\ mol\ H_2O)\left(\frac{18.02\ g}{mol}\right) = 900.\ g\ H_2O$

$(900.\ g)(2.26\ kJ/g) + (900.\ g)\left(\frac{4.184\ J}{g\ °C}\right)(100.0°C - 30.0°C) = 2.03 \times 10^6\ J + 2.64 \times 10^5\ J$

$= 2.29 \times 10^6\ J$

81. Energy to warm the ice from −10.0°C to 0°C

$(100.\ g)\left(\frac{2.01\ J}{g\ °C}\right)(10.0°C) = 2010\ J$

Energy to melt the ice at 0°C

$(100.\ g)(335\ J/g) = 33,500\ J$

Energy to heat the water from 0°C to 20.0°C

$(100.\ g)\left(\frac{4.184\ J}{g\ °C}\right)(20.0°C) = 8370\ J$

$E_{total} = 2010\ J + 33,500\ J + 8370\ J = 4.39 \times 10^4\ J = 43.9\ kJ$

82. $2\ H_2O \longrightarrow 2\ H_2 + O_2$

$(25.0\ L\ O_2)\left(\frac{1\ mol}{22.4\ L}\right)\left(\frac{2\ mol\ H_2O}{1\ mol\ O_2}\right)\left(\frac{18.02\ g}{mol}\right) = 40.2\ g\ H_2O$

83. 1.00 mol of liquid water has a density of 1.00 g/mL.

$$d = \frac{m}{V} \qquad V = \frac{m}{d} = \frac{18.02 \text{ g}}{1.00 \text{ g/mL}} = 18.0 \text{ mL}$$

1.00 mol of water vapor at STP has a volume of 22.4 L **(gas)**

84. (a) $2 \text{ K} + 2 \text{ H}_2\text{O} \longrightarrow 2 \text{ KOH} + \text{H}_2$

$$(1.00 \text{ mol K})\left(\frac{2 \text{ mol H}_2\text{O}}{2 \text{ mol K}}\right)\left(\frac{18.02 \text{ g}}{\text{mol}}\right) = 18.0 \text{ g H}_2\text{O}$$

(b) $\text{Ca} + 2 \text{ H}_2\text{O} \longrightarrow \text{Ca(OH)}_2 + \text{H}_2$

$$(1.00 \text{ mol Ca})\left(\frac{2 \text{ mol H}_2\text{O}}{1 \text{ mol Ca}}\right)\left(\frac{18.02 \text{ g}}{\text{mol}}\right) = 36.0 \text{ g H}_2\text{O}$$

(c) $2 \text{ Na} + 2 \text{ H}_2\text{O} \longrightarrow 2 \text{ NaOH} + \text{H}_2$

$$(1.00 \text{ g Na})\left(\frac{1 \text{ mol}}{22.99 \text{ g}}\right)\left(\frac{2 \text{ mol H}_2\text{O}}{2 \text{ mol Na}}\right)\left(\frac{18.02 \text{ g}}{\text{mol}}\right) = 0.784 \text{ g H}_2\text{O}$$

(d) $\text{SO}_3 + \text{H}_2\text{O} \longrightarrow \text{H}_2\text{SO}_4$

$$(1.00 \text{ mol SO}_3)\left(\frac{1 \text{ mol H}_2\text{O}}{1 \text{ mol SO}_3}\right)\left(\frac{18.02 \text{ g}}{\text{mol}}\right) = 18.0 \text{ g H}_2\text{O}$$

(e) $\text{MgO} + \text{H}_2\text{O} \longrightarrow \text{Mg(OH)}_2$

$$(1.00 \text{ g MgO})\left(\frac{1 \text{ mol}}{40.31 \text{ g}}\right)\left(\frac{1 \text{ mol H}_2\text{O}}{1 \text{ mol MgO}}\right)\left(\frac{18.02 \text{ g}}{\text{mol}}\right) = 0.447 \text{ g H}_2\text{O}$$

(f) $\text{N}_2\text{O}_5 + \text{H}_2\text{O} \longrightarrow 2 \text{ HNO}_3$

$$(1.00 \text{ g N}_2\text{O}_5)\left(\frac{1 \text{ mol}}{108.0 \text{ g}}\right)\left(\frac{1 \text{ mol H}_2\text{O}}{1 \text{ mol N}_2\text{O}_5}\right)\left(\frac{18.02 \text{ g}}{\text{mol}}\right) = 0.167 \text{ g H}_2\text{O}$$

85. $\left(\frac{1.00 \text{ mol H}_2\text{O}}{\text{day}}\right)\left(\frac{6.022 \times 10^{23} \text{ molecules}}{\text{mol}}\right)\left(\frac{1.00 \text{ day}}{24 \text{ hr}}\right)\left(\frac{1 \text{ hr}}{60 \text{ min}}\right)\left(\frac{1 \text{ min}}{60 \text{ s}}\right) = 6.97 \times 10^{18} \text{ molecules/s}$

86. Mass solution $-$ mass H_2O = mass H_2SO_4

(122 mL)(1.26 g/mL) $-$ (100. mL)(1.00 g/mL) = 54 g H_2SO_4

87. $2 \text{ H}_2 + \text{O}_2 \longrightarrow 2 \text{ H}_2\text{O}$

(a) $(80.0 \text{ mL H}_2)\left(\frac{1 \text{ mL O}_2}{2 \text{ mL H}_2}\right) = 40.0 \text{ mL O}_2$

Since 60.0 mL are available, oxygen remains.

(b) 60.0 mL $-$ 40.0 mL = 20.0 mL O_2 unreacted.

SOLUTIONS

1. (a) 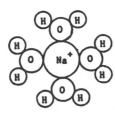 (b)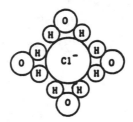

These diagrams are intended to illustrate the orientation of the water molecules about the ions, not the number of water molecules.

2. Reasonably soluble:

   (a) KOH   (b) $NiCl_2$ (d) $AgC_2H_3O_2$
   (e) Na_2CrO_4 (h) $CaCl_2$ (i) $Fe(NO_3)_3$

   Insoluble:

   (c) ZnS   (f) PbI_2 (g) $MgCO_3$ (j) $BaSO_4$

3. From Table 15.3, approximately 4.5 g of NaF would be soluble in 100 g of water at 50°C.

4. From Figure 15.3, solubilities in water at 25°C are:

   (a) KCl   35 g/100 g H_2O
   (b) $KClO_3$ 9 g/100 g H_2O
   (c) KNO_3 39 g/100 g H_2O

5. Potassium fluoride has a relatively high solubility when compared to lithium or sodium fluoride. For lithium and sodium halides, the order of solubility (in order of increasing solubilities) is:

   F^- Cl^- Br^- I^-

   For potassium halides, the order of increasing solubilities is:

   Cl^- Br^- F^- I^-

6. (a) $KClO_3$ at 60°C, 25 g   (c) Li_2SO_4 at 80°C, 31 g

   (b) HCl at 20°C, 72 g   (d) KNO_3 at 0°C, 14 g

7. KNO_3

8. A one molal solution in camphor will show a greater freezing point depression than a 2 molal solution in benzene.

$$\Delta t_f = \left(\frac{1 \text{ mol solute}}{\text{kg camphor}}\right)\left(\frac{40°C \text{ kg camphor}}{\text{mol solute}}\right) = 40°C$$

$$\Delta t_f = \left(\frac{2 \text{ mol solute}}{\text{kg benzene}}\right)\left(\frac{5.1°C \text{ kg benzene}}{\text{mol solute}}\right) = 10.2°C$$

9.

	1 cm	0.01 cm
Cube		
Volume	1 cm^3	$1 \times 10^{-6} \text{ cm}^3$
Number/1 cm cube	1	10^6
Area of face	1 cm^2	$1 \times 10^{-4} \text{ cm}^2$
Total surface area	6 cm^2	$6 \times 10^2 \text{ cm}^2$

$(1 \times 10^6 \text{ cubes})(6 \text{ faces/cube})(1 \times 10^{-4} \text{ cm}^2/\text{face}) = 6 \times 10^2 \text{ cm}^2$

10. $\dfrac{63 \text{ g NH}_4\text{Cl}}{150 \text{ g H}_2\text{O}} = \dfrac{42 \text{ g NH}_4\text{Cl}}{100 \text{ g H}_2\text{O}}$ From Figure 15.3, the solubility of

NH$_4$Cl in water is approximately 42 g/100 g H$_2$O at 30°C, 45 g/100 g H$_2$O at 40°C. Therefore, the solution of 63 g/150 g of water would be saturated at 10°C, 20°C, and 30°C. The solution would be unsaturated at 40°C and 50°C.

11. The dissolving process involves solvent molecules attaching to the solute ions or molecules. This rate decreases as more of the solvent molecules are already attached to solute molecules. As the solution becomes more saturated, the number of unused solvent molecules is much less. Also, the rate of recrystallization increases as the concentration of dissolved solute increases.

12. Yes. A volumetric flask is satisfactory for preparing normal solutions:

   (a)   Determine of the number of equivalents of solute required for the volume of solution.

   (b)   Measure out the corresponding number of grams of solute.

   (c)   Place the solute in the volumetric flask.

   (d)   Add water to dissolve the solute.

   (e)   Dilute to the appropriate mark with water and mix.

13. Because the concentration of water is greater in the thistle tube, the water will flow through the membrane from the thistle tube to the urea solution in the beaker. The solution level in the thistle tube will fall.

14. The two components of a solution are the solute and the solvent. The solute is dissolved into the solvent or is the least abundant component. The solvent is the dissolving agent or the most abundant component.

15. It is not always apparent which component in a solution is the solute. For example, in a solution composed of equal volumes of two liquids, the designation of solute and solvent would be simply a matter of preference on the part of the person making the designation.

16. The ions or molecules of a dissolved solute do not settle out because the individual particles are so small that the force of molecular collisions is large compared to the force of gravity.

17. Yes. It is possible to have one solid dissolved in another solid. Metal alloys are of this type. Atoms of one metal are dissolved among atoms of another metal.

18. Orange. The three reference solutions are KCl, $KMnO_4$, and $K_2Cr_2O_7$. They all contain K^+ ions in solution. The different colors must result from the different anions dissolved in the solutions: MnO_4^- (purple) and $Cr_2O_7^{2-}$ (orange). Therefore, it is predictable that the $Cr_2O_7^{2-}$ ion present in an aqueous solution of $Na_2Cr_2O_7$ will impart an orange color to the solution.

19. Carbon tetrachloride and benzene are both nonpolar molecules. There are no strong intermolecular forces between molecules of either substance or with each other, so they are miscible. Sodium chloride consists of ions strongly attracted to each other by electrical attractions. The carbon tetrachloride molecules, being nonpolar, have no strong forces to pull the ions apart, so sodium chloride is insoluble in carbon tetrachloride.

20. Coca Cola has two main characteristics, taste and fizz (carbonation). The carbonation is due to a dissolved gas, carbon dioxide. Since dissolved gases become less soluble as temperature increases, warm Coca Cola would be flat, with little to no carbonation. It would, therefore, be unappealing to most people.

21. Air is considered to be a solution because it is a homogeneous mixture of several substances and does not have a fixed composition.

22. A teaspoon of sugar would definitely dissolve more rapidly in 200 mL of hot coffee than in 200 mL of iced tea. The much greater thermal agitation of the hot coffee will help break the sugar molecules away from the undissolved solid and disperse them throughout the solution. Other solutes in coffee and tea would have no significant effect. The temperature difference is the critical factor.

23. The solubility of gases in liquids is greatly affected by the pressure of a gas above the liquid. The greater the pressure, the more soluble the gas. There is very little effect of pressure regarding the dissolution of solids in liquids.

24. For a given mass of solute, the smaller the particles, the faster the dissolution of the solute. This is due to the smaller particles having a greater surface area exposed to the dissolving action of the solvent.

25. In a saturated solution, the net rate of dissolution is zero. There is no further increase in the amount of dissolved solute, even though undissolved solute is continuously dissolving, because dissolved solute is continuously coming out of solution at a rate equal to the rate of dissolving.

26. When crystals of $AgNO_3$ and $NaCl$ are mixed, the contact is not intimate enough for the double displacement reaction to occur. When solutions of the two chemicals are mixed, the ions are free to move and come into intimate contact with each other, allowing the reaction to occur easily. The $AgCl$ formed is insoluble.

27. A 16 molar solution of nitric acid is a solution where the ratio of moles of solute (HNO_3) to liters of solution is 16:1. 16 moles HNO_3 per liter of solution.

28. The two solutions contain the same number of chloride ions. One liter of 1 M NaCl contains 1 mole of NaCl, therefore 1 mole of chloride ions. 0.5 liter of 1 M $MgCl_2$ contains 0.5 mole of $MgCl_2$ and 1 mole of chloride ions.

$$(0.5\ L)\left(\frac{1\ mol\ MgCl_2}{L}\right)\left(\frac{2\ mol\ Cl^-}{1\ mol\ MgCl_2}\right)\ =\ 1\ mol\ Cl^-$$

29. The champagne would spray out of the bottle all over the place. The rise in temperature and the increase in kinetic energy of the molecules by shaking both act to decrease the solubility of gas within the liquid. The pressure inside the bottle would be great. As the cork is popped, much of the gas would escape from the liquid very rapidly, causing the champagne to spray.

30. A supersaturated solution of $NaC_2H_3O_2$ may be prepared in the following sequence:

(a)   Determine the mass of $NaC_2H_3O_2$ necessary to saturate a specific amount of water at

room temperature.

(b)   Place a bit more $NaC_2H_3O_2$ in the water than the amount needed to saturate the solution.

(c)   Heat the solution until all the solid dissolves.

(d)   Cover the container and allow it to cool undisturbed. The cool solution, which sould contain no solid Na $C_2H_3O_2$, is supersaturated.

To test for supersaturation, add one small crystal of $NaC_2H_3O_2$ to the solution. Immediate crystallization is an indication that the solution was supersaturated.

31. A semipermeable membrane will allow water molecules to pass through in both directions. If it has pure water on one side and 10% sugar solutions on the other side of the membrane, there is a higher concentration of water molecules on the pure water side. More water molecules will pass from the pure water to the sugar solution.

32. The urea solution will have the greater osmotic pressure because it has 1.67 moles solute/kg H_2O, while the glucose solution has only 0.83 mole solute/kg H_2O.

33. A lettuce leaf immersed in salad dressing containing salt and vinegar will become limp and wilted as a result of osmosis. As the water inside the leaf flows into the dressing where the solute concentration is higher the leaf becomes limp from fluid loss. In water, osmosis proceeds in the opposite direction flowing into the lettuce leaf maintaining a high fluid content and crisp leaf.

34. The concentration of solutes is higher in seawater than in body fluids. The survivors who drank seawater suffered further dehydration from the transfer of water by osmosis from body tissues to the intestinal tract.

35. The correct statements are a, b, f, h, j, k, l, n, p, r, s, t, v, x, y, z

(c)   A solute cannot be removed from a solution by filtration.

(d)   Saturated solutions may or may not be concentrated solutions.

(e)   If a solution of sugar in water is allowed to stand undisturbed for a lengthy period, it will remain a solution.

(g)   Gases are usually more soluble in cold H_2O than in hot H_2O.

(i)   A solution that is 10% NaCl by mass always contains 10 g NaCl for each 100 g solution.

(m)   Dissolving 1 mole of NaCl in 1 liter (1000 g) of water will give a 1 molal solution.

(o)   When 100 mL of 0.200 M HCl is diluted to 200 mL volume by the addition of water, the resulting solution is 0.100 M and contains the same number of moles of HCl as were in the original solution.

(q)   100 mL of 0.1 N H_2SO_4 will neutralize the same volume of 0.1 M NaOH as 100 mL of 0.1 M HCl.

(u)   An aqueous solution that freezes below 0°C will have a normal boiling point above 100°C.

(w)   A solution of 1.00 mol of a nonionizable solute and 1000 g of water will freeze at –1.86°C and will boil at 100.5°C at atmospheric pressure.

36. "Dilute" and "concentrated" are rather general terms, which are not very useful, because they are so indefinite.   A solution that is termed dilute in one situation might well be considered concentrated under different circumstances.

37. A solution of H_2SO_4 that is 18 molar contains 18 moles of H_2SO_4 per liter of solution.   Since there can be 2 equivalents of H_2SO_4 per mole of H_2SO_4, a solution of 18 molar would contain 36 equivalents of H_2SO_4 per liter of solution.   Such a solution would be 36 normal.

38. The number of grams of NaCl in 750 mL of 5 molar solution is

$$(0.75 \text{ L})\left(\frac{5 \text{ mol NaCl}}{\text{L}}\right)\left(\frac{58.44 \text{ g}}{1 \text{ mol}}\right) = 200 \text{ g NaCl}$$

Dissolve the 200 g of NaCl in a minimum amount of water, then dilute the resulting solution to a final volume of 750 mL.

39. Ranking of the specified bases in descending order of the volume of each required to react with 1 liter of 1 M HCl.   The volume of each required to yield 1 mole of OH^- ion is shown.

(a)   1 M NaOH   1 liter

(d)   0.6 M $Ba(OH)_2$ 0.83 liter

(c)   2 M KOH   0.50 liter

(b)   1.5 M $Ca(OH)_2$ 0.33 liter

40. The boiling point of a liquid or solution is the temperature at which the vapor pressure of the liquid equals the pressure of the atmosphere.   Since a solution containing a nonvolatile solute has a lower vapor pressure than the pure solvent, the boiling point of the solution must be at a higher temperature than for the pure solvent.   This will result in the vapor pressure of the solution equaling the atmospheric pressure.

41. The freezing point is the temperature at which a liquid changes to a solid.   The vapor pressure of a solution is lower than that of a pure solute.   Therefore, the vapor pressure curve of the solution intersects the vapor pressure curve of the pure solid, at a temperature lower than the freezing point of the pure solvent.   At this point of intersection, the vapor pressure of the solution equals the vapor pressure of the pure solvent.

42. A glass filled with Seven-Up and crushed ice would be colder than a glass of water and crushed ice. The ice will keep the solution at its freezing point, and the Seven-Up will have a lower freezing point because of the dissolved solutes.

43. Water and ice are different phases of the same substance in equilibrium at the freezing point of water, 0℃. The presence of the methanol lowers the freezing point of the water. If the ratio of alcohol to water is high, the freezing point can be lowered an much as 10℃ or more.

44. Effectiveness in lowering the freezing point of 500. g water:

(a)   100. g (2.17 mol) of ethyl alcohol is more effective than 100. g (0.292 mol) of sucrose.

(b)   20.0 g (0.435 mol) of ethyl alcohol is more effective than 20.0 g (0.292 mol) of sucrose.

(c)   20.0 g (0.625 mol) of methyl alcohol is more effective than 20.0 g (0.435 mol) of ethyl alcohol.

45. 5 molal NaCl = 5 mol NaCl/kg H_2O; 5 molar NaCl = 5 mol NaCl/L of solution. The volume of the 5 molal solution will be larger than 1 liter (1 L H_2O + 5 mol NaCl). The volume of the 5 molar solution is exactly 1 L (5 mol NaCl + sufficient H_2O to produce 1 L of solution). The 5 molal solution is less concentrated than the 5 molar solution.

46. The normality of H_2SO_4 equals twice the molarity; N = 2 M.

47. Microencapsulation is a technique in which reactive chemicals are sealed in tiny capsules. The chemicals are released at the appropriate time for the reaction to occur.

48. When the paper on a scratch and sniff label is scratched or pulled open the fragrance is released into the air.

49. Timed released microencapsulated products are used for pesticides, medications, and neutralizer for contact lenses.

50. Three types of microencapsulation systems are:

1)   water diffuses through capsule forming a solution which diffuses out again. This can be used in medications to release the drug over a period of time.

2)   mechanical. This type of microcapsule can be disturbed by pressure as in carbonless paper receipts, or adhesive label.

3)   thermal. This type of system is found in flavorings in prepackaged mixes and release contents during cooking.

51. Mass percent calculations.

(a)   25.0 g NaBr + 100. g H_2O = 125 g solution

$$\left(\frac{25.0 \text{ g NaBr}}{125 \text{ g solution}}\right)(100) = 20.0\% \text{ NaBr}$$

(b)   1.20 g K_2SO_4 + 10.0 g H_2O = 11.2 g solution

$$\left(\frac{1.20 \text{ g } K_2SO_4}{11.2 \text{ g solution}}\right)(100) = 10.7\% \text{ } K_2SO_4$$

(c) $40.0 \text{ g Mg(NO}_3)_2 + 500. \text{ g H}_2\text{O} = 540. \text{ g solution}$

$\left(\dfrac{40.0 \text{ g Mg(NO}_3)_2}{540. \text{ g solution}}\right)(100) = 7.41\% \text{ Mg(NO}_3)_2$

52. A 12.5% $AgNO_3$ solution contains 12.5 g $AgNO_3$ per 100. g solution

(a) $(30.0 \text{ g AgNO}_3)\left(\dfrac{100. \text{ g solution}}{12.5 \text{ g AgNO}_3}\right) = 240. \text{ g solution}$

(b) $(0.400 \text{ mol AgNO}_3)\left(\dfrac{169.9 \text{ g}}{\text{mol}}\right)\left(\dfrac{100. \text{ g solution}}{12.5 \text{ g AgNO}_3}\right) = 544 \text{ g solution}$

53. Mass percent calculations.

(a) $60.0 \text{ g NaCl} + 200.0 \text{ g H}_2\text{O} = 260.0 \text{ g solution}$

$\left(\dfrac{60.0 \text{ g NaCl}}{260.0 \text{ g solution}}\right)(100) = 23.1\% \text{ NaCl}$

(b) $(0.25 \text{ mol HC}_2\text{H}_3\text{O}_2)\left(\dfrac{60.03 \text{ g}}{\text{mol}}\right) = 15 \text{ g HC}_2\text{H}_3\text{O}_2$

$(3.0 \text{ mol H}_2\text{O})\left(\dfrac{18.02 \text{ g}}{\text{mol}}\right) = 54 \text{ g H}_2\text{O}$

$\left(\dfrac{15 \text{ g HC}_2\text{H}_3\text{O}_2}{69 \text{ g solution}}\right)(100) = 22\% \text{ HC}_2\text{H}_3\text{O}_2$

(c) $1.0 \text{ molal solution of C}_6\text{H}_{12}\text{O}_6 = \left(\dfrac{1 \text{ mol C}_6\text{H}_{12}\text{O}_6}{1000. \text{ g H}_2\text{O}}\right)$

$(1.0 \text{ mol C}_6\text{H}_{12}\text{O}_6)\left(\dfrac{180.2 \text{ g}}{\text{mol}}\right) = 180 \text{ g C}_6\text{H}_{12}\text{O}_6$

$1000. \text{ g H}_2\text{O} + 180 \text{ g C}_6\text{H}_{12}\text{O}_6 = 1200 \text{ g solution}$

$\left(\dfrac{180 \text{ g C}_6\text{H}_{12}\text{O}_6}{1200 \text{ g solution}}\right)(100) = 15\% \text{ C}_6\text{H}_{12}\text{O}_6$

54. (a) $(65 \text{ g solution})\left(\dfrac{5.0 \text{ g KCl}}{100. \text{ g solution}}\right) = 3.3 \text{ g KCl}$

(b) $(250. \text{ g solution})\left(\dfrac{15.0 \text{ g K}_2\text{CrO}_4}{100. \text{ g solution}}\right) = 37.5 \text{ g K}_2\text{CrO}_4$

(c) $(100.0 \text{ g H}_2\text{O})\left(\dfrac{6.0 \text{ g NaHCO}_3}{94.0 \text{ g H}_2\text{O}}\right) = 6.4 \text{ g NaHCO}_3$

55. $(25.0 \text{ g KCl})\left(\dfrac{100. \text{ g solution}}{5.50 \text{ g KCl}}\right) = 455 \text{ g solution}$

Alternate solution:

$\left(\dfrac{25.0 \text{ g KCl}}{x}\right) = \left(\dfrac{5.50 \text{ g KCl}}{100. \text{ g solution}}\right)$

$x = 455 \text{ g solution}$

56. (a) $(500. \text{ g solution})\left(\dfrac{0.90 \text{ g NaCl}}{100. \text{ g solution}}\right) = 4.5 \text{ g NaCl}$

(b) $(4.5 \text{ g NaCl})\left(\dfrac{91 \text{ g H}_2\text{O}}{9.0 \text{ g NaCl}}\right) = 46 \text{ g H}_2\text{O}$ needed to produce a solution that is 9.0% NaCl.

500. g solution $- 4.5$ g NaCl $= 495$ g H_2O

495 g $\text{H}_2\text{O} - 46$ g $\text{H}_2\text{O} = 449$ g H_2O must be evaporated.

57. From Figure 15.3, the solubility of KNO_3 in H_2O at 20°C is 32 g per 100 g H_2O.

$(50.0 \text{ g KNO}_3)\left(\dfrac{100. \text{ g H}_2\text{O}}{32.0 \text{ g KNO}_3}\right) = 156 \text{ g H}_2\text{O}$ to produce a saturated solution.

175 g $\text{H}_2\text{O} - 156$ g $\text{H}_2\text{O} = 19$ g H_2O must be evaporated.

58. Mass - volume percent.

(a) $\left(\dfrac{22.0 \text{ g CH}_3\text{OH}}{100. \text{ mL solution}}\right)(100) = 22.0\% \text{ CH}_3\text{OH}$

(b) $\left(\dfrac{4.20 \text{ g NaCl}}{12.5 \text{ mL solution}}\right)(100) = 33.6 \% \text{ NaCl}$

59. Volume percent.

(a) $\left(\dfrac{10.0 \text{ mL CH}_3\text{OH}}{40.0 \text{ mL solution}}\right)(100) = 25.0\% \text{ CH}_3\text{OH}$

(b) $\left(\dfrac{2.0 \text{ mL CCl}_4}{9.0 \text{ mL solution}}\right)(100) = 22\% \text{ CCl}_4$

60. $\dfrac{150 \text{ mL alcohol}}{x} = \dfrac{70. \text{ mL alcohol}}{100. \text{ mL solution}}$

$x = 2.1 \times 10^2 \text{ mL solution}$

61. (a) $(1.00 \text{ L solution})\left(\dfrac{1000 \text{ mL solution}}{\text{L solution}}\right)\left(\dfrac{1.21 \text{ g}}{\text{mL}}\right)\left(\dfrac{35.0 \text{ g HNO}_3}{100 \text{ g solution}}\right) = 424 \text{ g HNO}_3$

(b) $(500. \text{ g HNO}_3)\left(\dfrac{1000 \text{ mL}}{424 \text{ g HNO}_3}\right)\left(\dfrac{1.00 \text{ L}}{1000 \text{ mL}}\right) = 1.18 \text{ L}$

62. Molarity problems ($M = \dfrac{\text{mol}}{\text{L}}$)

(a) $\left(\dfrac{0.10 \text{ mol}}{250 \text{ mL}}\right)\left(\dfrac{1000 \text{ mL}}{\text{L}}\right) = 0.40 \text{ M}$

(b) $\left(\dfrac{2.5 \text{ mol NaCl}}{0.650 \text{ L}}\right) = 3.8 \text{ M NaCl}$

(c) $\left(\dfrac{0.025 \text{ mol HCl}}{10. \text{ mL}}\right)\left(\dfrac{1000 \text{ mL}}{\text{L}}\right) = 2.5 \text{ M HCl}$

(d) $\left(\dfrac{0.35 \text{ mol BaCl}_2 \cdot 2 \text{ H}_2\text{O}}{593 \text{ mL}}\right)\left(\dfrac{1000 \text{ mL}}{\text{L}}\right) = 0.59 \text{ M BaCl}_2 \cdot 2 \text{ H}_2\text{O}$

63. (a) $\left(\dfrac{53.0 \text{ g Na}_2\text{CrO}_4}{1.00 \text{ L}}\right)\left(\dfrac{1 \text{ mol}}{162.0 \text{ g}}\right) = 0.327 \text{ M Na}_2\text{CrO}_4$

(b) $\left(\dfrac{260 \text{ g C}_6\text{H}_{12}\text{O}_6}{800. \text{ mL}}\right)\left(\dfrac{1000 \text{ mL}}{\text{L}}\right)\left(\dfrac{1 \text{ mol}}{180.2 \text{ g}}\right) = 1.8 \text{ M C}_6\text{H}_{12}\text{O}_6$

(c) $\left(\dfrac{1.50 \text{ g Al}_2(\text{SO}_4)_3}{2.00 \text{ L}}\right)\left(\dfrac{1 \text{ mol}}{342.1 \text{ g}}\right) = 2.19 \times 10^{-3} \text{ M Al}_2(\text{SO}_4)_3$

(d) $\left(\dfrac{0.0282 \text{ g Ca(NO}_3)_2}{1.00 \text{ mL}}\right)\left(\dfrac{1000 \text{ mL}}{\text{L}}\right)\left(\dfrac{1 \text{ mol}}{164.1 \text{ g}}\right) = 0.172 \text{ M Ca(NO}_3)_2$

64. Molarity $= \dfrac{\text{mol solute}}{\text{L solution}}$

(a) $(40.0 \text{ L})\left(\dfrac{1.0 \text{ mol LiCl}}{1 \text{ L}}\right) = 40. \text{ mol LiCl}$

(b) $(25.0 \text{ mL})\left(\dfrac{1 \text{ L}}{1000 \text{ mL}}\right)\left(\dfrac{3.00 \text{ mol H}_2\text{SO}_4}{\text{L}}\right) = 0.0750 \text{ mol H}_2\text{SO}_4$

(c) $(349 \text{ mL})\left(\dfrac{1 \text{ L}}{1000 \text{ mL}}\right)\left(\dfrac{0.0010 \text{ mol NaOH}}{\text{L}}\right) = 3.5 \times 10^{-4} \text{ mol NaOH}$

(d) $(5000. \text{ mL})\left(\dfrac{1 \text{ L}}{1000 \text{ mL}}\right)\left(\dfrac{3.1 \text{ mol CoCl}_2}{\text{L}}\right) = 16 \text{ mol CoCl}_2$

65. (a) $(150 \text{ L})\left(\dfrac{1.0 \text{ mol NaCl}}{\text{L}}\right)\left(\dfrac{58.44 \text{ g}}{\text{mol}}\right) = 8.8 \times 10^3 \text{ g NaCl}$

(b) $(0.035 \text{ L})\left(\dfrac{10.0 \text{ mol HCl}}{\text{L}}\right)\left(\dfrac{36.46 \text{ g}}{\text{mol}}\right) = 13 \text{ g HCl}$

(c) $(260 \text{ mL})\left(\dfrac{18 \text{ mol H}_2\text{SO}_4}{1000 \text{ mL}}\right)\left(\dfrac{98.08 \text{ g}}{\text{mol}}\right) = 4.6 \times 10^2 \text{ g H}_2\text{SO}_4$

(d) $(8.00 \text{ mL})\left(\dfrac{1 \text{ L}}{1000 \text{ mL}}\right)\left(\dfrac{8.00 \text{ mol Na}_2\text{C}_2\text{O}_4}{\text{L}}\right)\left(\dfrac{134.0 \text{ g}}{\text{mol}}\right) = 8.58 \text{ g Na}_2\text{C}_2\text{O}_4$

66. (a) $(0.430 \text{ mol})\left(\dfrac{1 \text{ L}}{0.256 \text{ mol}}\right)\left(\dfrac{1000 \text{ mL}}{\text{L}}\right) = 1.68 \times 10^3 \text{ mL}$

(b) $(10.0 \text{ mol})\left(\dfrac{1 \text{ L}}{0.256 \text{ mol}}\right)\left(\dfrac{1000 \text{ mL}}{\text{L}}\right) = 3.91 \times 10^4 \text{ mL}$

(c) $(20.0 \text{ g})\left(\dfrac{1 \text{ mol}}{74.55 \text{ g}}\right)\left(\dfrac{1 \text{ L}}{0.256 \text{ mol}}\right)\left(\dfrac{1000 \text{ mL}}{\text{L}}\right) = 1.05 \times 10^3 \text{ mL}$

(d) $(71.0 \text{ g Cl}^-)\left(\dfrac{1 \text{ mol}}{35.45 \text{ g}}\right)\left(\dfrac{1 \text{ mol KCl}}{1 \text{ mol Cl}^-}\right)\left(\dfrac{1 \text{ L}}{0.256 \text{ mol KCl}}\right)\left(\dfrac{1000 \text{ mL}}{\text{L}}\right) = 7.82 \times 10^3 \text{ mL}$

67. Assume 1.00 L of solution.

$$\left(\dfrac{1000 \text{ mL solution}}{\text{L}}\right)\left(\dfrac{1.21 \text{ g}}{\text{mL}}\right)\left(\dfrac{35.0 \text{ g HNO}_3}{100. \text{ g solution}}\right)\left(\dfrac{1 \text{ mol}}{63.02 \text{ g}}\right) = 6.72 \text{ M HNO}_3$$

68. Dilution problem

$$V_1M_1 = V_2M_2$$

(a) $V_1 = 200.\text{ mL}$ $\qquad$ $V_2 = 400.\text{ mL}$

$\quad$ $M_1 = 12\text{ M}$ $\qquad\qquad$ $M_2 = ?$

$\quad$ $(200.\text{ mL})(12\text{ M}) = (400.\text{ mL})(M_2)$

$\quad$ $M_2 = \dfrac{(200.\text{ mL})(12\text{ M})}{400.\text{ mL}} = 6.0\text{ M HCl}$

(b) $V_1 = 60.0\text{ mL}$ $\qquad$ $V_2 = 560.\text{ mL}$

$\quad$ $M_1 = 0.60\text{ M}$ $\qquad\qquad$ $M_2 = ?$

$\quad$ $(60.0\text{ mL})(0.60\text{ M}) = (560.\text{ mL})(M_2)$

$\quad$ $M_2 = \dfrac{(60.0\text{ mL})(0.60\text{ M})}{560.\text{ mL}} = 0.064\text{ M ZnSO}_4$

(c) First calculate the moles of HCl in each solution. Then calculate the molarity.

$\quad$ $(100.\text{ mL})\left(\dfrac{1\text{ L}}{1000\text{ mL}}\right)\left(\dfrac{1.0\text{ mol}}{\text{L}}\right) = 0.10\text{ mol HCl}$

$\quad$ $(150.\text{ mL})\left(\dfrac{1\text{ L}}{1000\text{ mL}}\right)\left(\dfrac{2.0\text{ mol}}{\text{L}}\right) = 0.30\text{ mol HCl}$

$\quad$ Total mol $= 0.40\text{ mol HCl}$

$\quad$ Total volume $= 100.\text{ mL} + 150.\text{ mL} = 250.\text{ mL}$ (0.250 L)

$\quad$ $\dfrac{0.40\text{ mol HCl}}{0.250\text{ L}} = 1.6\text{ M HCl}$

69. $V_1M_1 = V_2M_2$

(a) $(V_1)(12\text{ M}) = (400.\text{ mL})(6.0\text{ M})$

$\quad$ $V_1 = \dfrac{(400.\text{ mL})(6.0\text{ M})}{12\text{ M}} = 2.0 \times 10^2\text{ mL 12 M HCl}$

(b) $(V_1)(15\text{ M}) = (50.\text{ mL})(6.0\text{ M})$

$\quad$ $V_1 = \dfrac{(50.\text{ mL})(6.0\text{ M})}{15\text{ M}} = 20.\text{ mL 15 M NH}_3$

(c) $(V_1)(16\text{ M}) = (100.\text{ mL})(2.5\text{ M})$

$\quad$ $V_1 = \dfrac{(100.\text{ mL})(2.5\text{ M})}{16\text{ M}} = 16\text{ mL 16 M HNO}_3$

(d) $10.0\text{ N H}_2\text{SO}_4 = \left(\dfrac{10.0\text{ equivalents H}_2\text{SO}_4}{\text{L}}\right)\left(\dfrac{1\text{ mol}}{2\text{ equivalents}}\right) = 5.00\text{ M H}_2\text{SO}_4$

$\quad$ $(V_1)(18\text{ M}) = (250\text{ mL})(5.00\text{ M})$

$\quad$ $V_1 = \dfrac{(250\text{ mL})(5.00\text{ M})}{18\text{ M}} = 69\text{ mL 18 M H}_2\text{SO}_4$

70. First calculate the molarity of the solution

$$\left(\frac{80.0 \text{ g H}_2\text{SO}_4}{500. \text{ mL}}\right)\left(\frac{1000 \text{ mL}}{\text{L}}\right)\left(\frac{1 \text{ mol}}{98.08 \text{ g}}\right) = 1.63 \text{ M H}_2\text{SO}_4$$

$$M_1V_1 = M_2V_2$$

$$(1.63 \text{ M})(500. \text{ mL}) = (0.10 \text{ M})(V_2)$$

$$V_2 = \frac{(1.63 \text{ M})(500. \text{ mL})}{0.10 \text{ M}} = 8.2 \times 10^3 \text{ mL} = 8.2 \text{ L}$$

71. $(0.250 \text{ L})\left(\frac{0.75 \text{ mol}}{\text{L}}\right) = 0.19 \text{ mol H}_2\text{SO}_4$

(a)   Final volume after mixing

$$250. \text{ mL} + 150. \text{ mL} = 400. \text{ mL} = 0.400 \text{ L}$$

$$\frac{0.19 \text{ mol H}_2\text{SO}_4}{0.400 \text{ L}} = 0.48 \text{ M H}_2\text{SO}_4$$

(b) $(250. \text{ mL})\left(\frac{1 \text{ L}}{1000 \text{ mL}}\right)\left(\frac{0.70 \text{ mol H}_2\text{SO}_4}{\text{L}}\right) = 0.18 \text{ mol H}_2\text{SO}_4$

Total moles $= 0.19 \text{ mol} + 0.18 \text{ mol} = 0.37 \text{ mol H}_2\text{SO}_4$

Final volume $= 250. \text{ mL} + 250. \text{ mL} = 500. \text{ mL} = 0.500 \text{ L}$

$$\frac{0.37 \text{ mol H}_2\text{SO}_4}{0.500 \text{ L}} = 0.74 \text{ M H}_2\text{SO}_4$$

(c) $(400. \text{ mL})\left(\frac{1 \text{ L}}{1000 \text{ mL}}\right)\left(\frac{2.50 \text{ mol H}_2\text{SO}_4}{\text{L}}\right) = 1.00 \text{ mol H}_2\text{SO}_4$

Total moles $= 0.19 \text{ mol} + 1.00 \text{ mol} = 1.19 \text{ mol H}_2\text{SO}_4$

Final volume $= 250. \text{ mL} + 400. \text{ mL} = 650. \text{ mL} = 0.650 \text{ L}$

$$\frac{1.19 \text{ mol H}_2\text{SO}_4}{0.650 \text{ L}} = 1.83 \text{ M H}_2\text{SO}_4$$

72. Note that the problem asks for the volume of water to be added, not the final volume.

$$(300. \text{ mL})(1.40 \text{ M}) = (V_2)(0.500 \text{ M})$$

$$V_2 = \frac{(300. \text{ mL})(1.40 \text{ M})}{0.500 \text{ M}} = 840. \text{ mL (final volume)}$$

$$840. \text{ mL} - 300. \text{ mL} = 540. \text{ mL water to be added}$$

73. $(10.0 \text{ mL})(16 \text{ M}) = (500. \text{ mL})(M_2)$

$$M_2 = \frac{(10.0 \text{ mL})(16 \text{ M})}{500. \text{ mL}} = 0.32 \text{ M HNO}_3$$

74. $(V_1)(5.00 \text{ M}) = (250 \text{ mL})(0.625 \text{ M})$

$$V_1 = \frac{(250 \text{ mL})(0.625 \text{ M})}{5.00 \text{ M}} = 31 \text{ mL } 5.00 \text{ M KOH}$$

To make 250. mL of 0.625 M KOH, take 31 mL of 5.00 M KOH and dilute with water to a volume of 250. mL.

75. $BaCl_2 + K_2CrO_4 \longrightarrow BaCrO_4 + 2 \text{ KCl}$

mL $BaCl_2 \longrightarrow$ mol $BaCl_2 \longrightarrow$ mol $BaCrO_4 \longrightarrow$ g $BaCrO_4$

(a) $(100. \text{ mL } BaCl_2)\left(\frac{0.300 \text{ mol}}{1000 \text{ mL}}\right)\left(\frac{1 \text{ mol } BaCrO_4}{1 \text{ mol } BaCl_2}\right)\left(\frac{253.3 \text{ g}}{\text{mol}}\right) = 7.60 \text{ g } BaCrO_4$

(b) $(50.0 \text{ mL } K_2CrO_4)\left(\frac{0.300 \text{ mol}}{1000 \text{ mL}}\right)\left(\frac{1 \text{ mol } BaCl_2}{1 \text{ mol } K_2CrO_4}\right)\left(\frac{1000 \text{ mL}}{1.0 \text{ mol}}\right) = 15 \text{ mL of } 1.0 \text{ mol } BaCl_2$

76. $3 \text{ MgCl}_2 + 2 \text{ Na}_3PO_4 \longrightarrow Mg_3(PO_4)_2 + 6 \text{ NaCl}$

(a) mL $MgCl_2 \longrightarrow$ mol $MgCl_2 \longrightarrow$ mol $Na_3PO_4 \longrightarrow$ mL Na_3PO_4

$(50.0 \text{ mL } MgCl_2)\left(\frac{0.250 \text{ mol}}{1000 \text{ mL}}\right)\left(\frac{2 \text{ mol } Na_3PO_4}{3 \text{ mol } MgCl_2}\right)\left(\frac{1000 \text{ mL}}{0.250 \text{ mol}}\right) = 33.3 \text{ mL of } 0.250 \text{ M } Na_3PO_4$

(b) mL $MgCl_2 \rightarrow$ mol $MgCl_2 \rightarrow$ mol $Mg_3(PO_4)_2 \rightarrow$ g $Mg_3(PO_4)_2$

$(50.0 \text{ mL } MgCl_2)\left(\frac{0.250 \text{ mol}}{1000 \text{ mL}}\right)\left(\frac{1 \text{ mol } Mg_3(PO_4)_2}{3 \text{ mol } MgCl_2}\right)\left(\frac{262.9 \text{ g}}{\text{mol}}\right) = 1.10 \text{ g } Mg_3(PO_4)_2$

77. $Mg + 2 \text{ HCl} \longrightarrow MgCl_2 + H_2\uparrow$

(a) mL HCl $\longrightarrow$ mol HCl $\longrightarrow$ mol H_2

$(200. \text{ mL HCl})\left(\frac{3.00 \text{ mol}}{1000 \text{ mL}}\right)\left(\frac{1 \text{ mol } H_2}{2 \text{ mol HCl}}\right) = 0.300 \text{ mol } H_2$

(b) $PV = nRT$

$P = \left(\frac{720}{760}\right)\text{atm} = 0.947 \text{ atm}$

$T = 27°C = 300. \text{ K}$

$n = 0.300 \text{ mol}$

$V = \frac{nRT}{P} = \frac{(0.300 \text{ mol})(0.0821 \text{ L atm/mol K})(300. \text{ K})}{0.947 \text{ atm}} = 7.80 \text{ L } H_2$

78. $Mg + 2 \text{ HCl} \longrightarrow MgCl_2 + H_2\uparrow$

L $H_2 \longrightarrow$ mol $H_2 \longrightarrow$ mol HCl $\longrightarrow$ M HCl

$(3.50 \text{ L } H_2)\left(\frac{1 \text{ mol}}{22.4 \text{ L}}\right)\left(\frac{2 \text{ mol HCl}}{1 \text{ mol } H_2}\right)\left(\frac{1}{0.150 \text{ L}}\right) = 2.08 \text{ M HCl}$

79. The balanced equation is

$$6\ FeCl_2 + K_2Cr_2O_7 + 14\ HCl \rightarrow 6\ FeCl_3 + 2\ CrCl_3 + 2\ KCl + 7H_2O$$

(a) $(2.0\ \text{mol}\ FeCl_2)\left(\dfrac{2\ \text{mol}\ KCl}{6\ \text{mol}\ FeCl_2}\right) = 0.67\ \text{mol}\ KCl$

(b) $(1.0\ \text{mol}\ FeCl_2)\left(\dfrac{2\ \text{mol}\ CrCl_3}{6\ \text{mol}\ FeCl_2}\right) = 0.33\ \text{mol}\ CrCl_3$

(c) $(0.050\ \text{mol}\ K_2Cr_2O_7)\left(\dfrac{6\ \text{mol}\ FeCl_2}{1\ \text{mol}\ K_2Cr_2O_7}\right) = 0.30\ \text{mol}\ FeCl_2$

(d) $(0.025\ \text{mol}\ FeCl_2)\left(\dfrac{1\ \text{mol}\ K_2Cr_2O_7}{6\ \text{mol}\ FeCl_2}\right)\left(\dfrac{1000\ \text{mL}}{0.060\ \text{mol}}\right) = 69\ \text{mL of } 0.060\ M\ K_2Cr_2O_7$

(e) $(15.0\ \text{mL}\ FeCl_2)\left(\dfrac{6.0\ \text{mol}}{1000\ \text{mL}}\right)\left(\dfrac{14\ \text{mol}\ HCl}{6\ \text{mol}\ FeCl_2}\right)\left(\dfrac{1000\ \text{mL}}{6\ \text{mol}}\right) = 35\ \text{mL of } 6.0\ M\ HCl$

80. $2\ KMnO_4 + 16\ HCl \longrightarrow 2\ MnCl_2 + 5\ Cl_2 + 8\ H_2O + 2\ KCl$

(a) $(0.50\ \text{mol}\ KMnO_4)\left(\dfrac{5\ \text{mol}\ Cl_2}{2\ \text{mol}\ KMnO_4}\right) = 1.3\ \text{mol}\ Cl_2$

(b) $(1.0\ L\ KMnO_4)\left(\dfrac{2.0\ \text{mol}}{L}\right)\left(\dfrac{16\ \text{mol}\ HCl}{2\ \text{mol}\ KMnO_4}\right) = 16\ \text{mol}\ HCl$

(c) $(200.\ \text{mL}\ KMnO_4)\left(\dfrac{0.50\ \text{mol}}{1000\ \text{mL}}\right)\left(\dfrac{16\ \text{mol}\ HCl}{2\ \text{mol}\ KMnO_4}\right)\left(\dfrac{1000\ \text{mL}}{6.0\ \text{mol}}\right) = 1.3 \times 10^2\ \text{mL of } 6\ M\ HCl$

(d) $(75.0\ \text{mL}\ HCl)\left(\dfrac{6.0\ \text{mol}}{1000\ \text{mL}}\right)\left(\dfrac{5\ \text{mol}\ Cl_2}{16\ \text{mol}\ HCl}\right)\left(\dfrac{22.4\ L}{\text{mol}}\right) = 3.2\ L\ Cl_2$

81. For an acid, the number of hydrogen ions reacting per formula unit of acid equals the number of equivalents per mole of acid. For a base, the number of hydroxide ions reacting per formula unit of base equals the number of equivalents per mole of base.

(a) HCl $\left(\dfrac{1\ \text{mol}\ HCl}{eq}\right)\left(\dfrac{36.46\ g}{\text{mol}}\right) = \left(\dfrac{36.46\ g\ HCl}{eq}\right)$

NaOH $\left(\dfrac{1\ \text{mol}\ NaOH}{eq}\right)\left(\dfrac{40.00\ g}{\text{mol}}\right) = \left(\dfrac{40.00\ g\ NaOH}{eq}\right)$

(b) HCl $\left(\dfrac{1\ \text{mol}\ HCl}{eq}\right)\left(\dfrac{36.46\ g}{\text{mol}}\right) = \left(\dfrac{36.46\ g\ HCl}{eq}\right)$

Ba(OH)$_2$ $\left(\dfrac{1\ \text{mol}\ Ba(OH)_2}{2\ eq}\right)\left(\dfrac{171.3\ g}{\text{mol}}\right) = \left(\dfrac{85.65\ g\ Ba(OH)_2}{eq}\right)$

(c) H_2SO_4 $\left(\dfrac{1\ \text{mol}\ H_2SO_4}{2\ eq}\right)\left(\dfrac{98.08\ g}{\text{mol}}\right) = \left(\dfrac{49.04\ g\ H_2SO_4}{eq}\right)$

Ca(OH)$_2$ $\left(\dfrac{1\ \text{mol}\ Ca(OH)_2}{2\ eq}\right)\left(\dfrac{74.10\ g}{\text{mol}}\right) = \left(\dfrac{37.05\ g\ Ca(OH)_2}{eq}\right)$

(d) H_2SO_4 $\left(\dfrac{1\text{ mol }H_2SO_4}{2\text{ eq}}\right)\left(\dfrac{98.08\text{ g}}{\text{mol}}\right) = \left(\dfrac{49.04\text{ g }H_2SO_4}{\text{eq}}\right)$

KOH $\left(\dfrac{1\text{ mol KOH}}{1\text{ eq}}\right)\left(\dfrac{56.11\text{ g}}{\text{mol}}\right) = \left(\dfrac{56.11\text{ g KOH}}{\text{eq}}\right)$

(e) H_3PO_4 $\left(\dfrac{1\text{ mol }H_3PO_4}{2\text{ eq}}\right)\left(\dfrac{97.99\text{ g}}{\text{mol}}\right) = \left(\dfrac{49.00\text{ g }H_3PO_4}{\text{eq}}\right)$

LiOH $\left(\dfrac{1\text{ mol LiOH}}{1\text{ eq}}\right)\left(\dfrac{23.95\text{ g}}{\text{mol}}\right) = \left(\dfrac{23.95\text{ g LiOH}}{\text{eq}}\right)$

82. Normality

(a) 4.0 M HCl $\quad\left(\dfrac{4.0\text{ mol}}{L}\right)\left(\dfrac{1\text{ eq}}{\text{mol}}\right) = \left(\dfrac{4.0\text{ eq}}{L}\right) = 4.0$ N HCl

(b) 0.243 M HNO_3 $\quad\left(\dfrac{0.243\text{ mol}}{L}\right)\left(\dfrac{1\text{ eq}}{\text{mol}}\right) = \left(\dfrac{0.243\text{ eq}}{L}\right) = 0.243$ N HNO_3

(c) 3.0 M H_2SO_4 $\quad\left(\dfrac{3.0\text{ mol}}{L}\right)\left(\dfrac{2\text{ eq}}{\text{mol}}\right) = \left(\dfrac{6.0\text{ eq}}{L}\right) = 6.0$ N H_2SO_4

(d) 1.85 M H_3PO_4 $\quad\left(\dfrac{1.85\text{ mol}}{L}\right)\left(\dfrac{3\text{ eq}}{\text{mol}}\right) = \left(\dfrac{5.55\text{ eq}}{L}\right) = 5.55$ N H_3PO_4

(e) 0.250 M $HC_2H_3O_2$ $\quad\left(\dfrac{0.250\text{ mol}}{L}\right)\left(\dfrac{1\text{ eq}}{\text{mol}}\right) = \left(\dfrac{0.250\text{ eq}}{L}\right) = 0.250$ N $HC_2H_3O_2$

83. Equivalent mass of $Ca(OH)_2 = \left(\dfrac{74.10\text{ g}}{\text{mol}}\right)\left(\dfrac{1\text{ mol}}{2\text{ eq}}\right) = \left(\dfrac{37.05\text{ g}}{\text{eq}}\right)$

Equivalents acid = equivalents base

$(N_A)(0.03626\text{ L}) = (2.50\text{ g }Ca(OH)_2)\left(\dfrac{1\text{ eq}}{37.05\text{ g}}\right)$

$N_A = \left(\dfrac{1.86\text{ g}}{\text{eq}}\right) = 1.86$ N H_2SO_4

84. $(12.0\text{ g }Mg(OH)_2)\left(\dfrac{1\text{ eq}}{29.16\text{ g }Mg(OH)_2}\right) = 0.411$ eq $Mg(OH)_2$

$(10.0\text{ g }Al(OH)_3)\left(\dfrac{1\text{ eq}}{26.00\text{ g }Al(OH)_3}\right) = 0.385$ eq $Al(OH)_3$

1 equivalent of base will neutralize 1 equivalent of acid. Therefore, 12.0 g $Mg(OH)_2$ will neutralize more stomach acid than 10.0 g $Al(OH)_3$.

85. $N_A V_A = N_B V_B$

(a) $(0.1254\text{ N})(20.22\text{ mL}) = (0.2250\text{ N})(V_B)$

$V_B = \dfrac{(0.1254\text{ N})(20.22\text{ mL})}{0.2550\text{ N}} = 9.943$ mL NaOH

(b) $(0.1246 \text{ N})(14.86 \text{ mL}) = (0.2550 \text{ N})(V_B)$

$$V_B = \frac{(0.1246 \text{ N})(14.86 \text{ mL})}{0.2550 \text{ N}} = 7.261 \text{ mL NaOH}$$

(c) $0.1430 \text{ M } H_2SO_4 = 0.2860 \text{ N } H_2SO_4$

(there are 2 equivalents of H in 1 mole of H_2SO_4)
$(0.2860 \text{ N})(18.00 \text{ mL}) = (0.2550 \text{ N})(V_B)$

$$V_B = \frac{(0.2860 \text{ N})(18.00 \text{ mL})}{0.2550 \text{ N}} = 20.19 \text{ mL NaOH}$$

86. Molality $= m = \dfrac{\text{mol solute}}{\text{kg solvent}}$

(a) $\left(\dfrac{14.0 \text{ g } CH_3OH}{100. \text{ g } H_2O}\right)\left(\dfrac{1000 \text{ g}}{\text{kg}}\right)\left(\dfrac{1 \text{ mol}}{32.04 \text{ g}}\right) = \left(\dfrac{4.37 \text{ mol } CH_3OH}{\text{kg } H_2O}\right) = 4.37 \ m \ \ CH_3OH$

(b) $\left(\dfrac{2.50 \text{ mol } C_6H_6}{250 \text{ g } CCl_4}\right)\left(\dfrac{1000 \text{ g}}{\text{kg}}\right) = \left(\dfrac{10.0 \text{ mol } C_6H_6}{\text{kg } CCl_4}\right) = 10 \ m \ C_6H_6$

(c) $\left(\dfrac{1.0 \text{ g } C_6H_{12}O_6}{1.0 \text{ g } H_2O}\right)\left(\dfrac{1000 \text{ g}}{\text{kg}}\right)\left(\dfrac{1 \text{ mol}}{180.2 \text{ g}}\right) = \left(\dfrac{5.6 \text{ mol } C_6H_{12}O_6}{\text{kg } H_2O}\right) = 5.5 \ m \ C_6H_{12}O_6$

87. (a) With equal masses of CH_3OH and C_2H_5OH, the substance with the lower molar mass will represent more moles of solute in solution. Therefore, the CH_3OH will be more effective than C_2H_5OH as an antifreeze.

(b) Equal molal solutions will lower the freezing point of the solution by the same amount.

88. Calculate molarity and molality. Assume 1000 mL of solution to calculate the amounts of H_2SO_4 and H_2O in the solution.

$$(1000 \text{ mL solution})\left(\frac{1.29 \text{ g}}{\text{mL}}\right) = 1.29 \times 10^3 \text{ g solution}$$

$$(1.29 \times 10^3 \text{ g solution})\left(\frac{38 \text{ g } H_2SO_4}{100 \text{ g solution}}\right) = 4.9 \times 10^2 \text{ g } H_2SO_4$$

$$1.29 \times 10^3 \text{ g solution} - 4.9 \times 10^2 \text{ g } H_2SO_4 = 8.0 \times 10^2 \text{ g } H_2O \text{ in the solution}$$

$$m = \left(\frac{490 \text{ g } H_2SO_4}{8.0 \times 10^2 \text{ g } H_2O}\right)\left(\frac{1000 \text{ g}}{\text{kg}}\right)\left(\frac{1 \text{ mol}}{98.08 \text{ g}}\right) = 6.2 \ m \ H_2SO_4$$

$$M = \left(\frac{4.9 \times 10^2 \text{ g } H_2SO_4}{L}\right)\left(\frac{1 \text{ mol}}{98.08 \text{ g}}\right) = 5.0 \text{ M } H_2SO_4$$

89. $1.00 \text{ lb} = 454 \text{ g sugar } (C_{12}H_{22}O_{11})$

$$(4.00 \text{ lb } H_2O)\left(\frac{454 \text{ g}}{\text{lb}}\right) = 1.82 \times 10^3 \text{ g } H_2O \ \ (1.82 \text{ kg } H_2O)$$

$$(454 \text{ g } C_{12}H_{22}O_{11})\left(\frac{1 \text{ mol}}{342.3 \text{ g}}\right) = 1.33 \text{ mol } C_{12}H_{22}O_{11}$$

K_f (for H_2O) = 1.86°C kg solvent/mol solute

$$\Delta t_f = mK_f = \left(\frac{1.33 \text{ mol } C_{12}H_{22}O_{11}}{1.82 \text{ kg } H_2O}\right)\left(\frac{1.86 \text{ °C kg } H_2O}{\text{mol } C_{12}H_{22}O_{11}}\right) = 1.36\text{°C}$$

Freezing point of solution = 0°C − 1.36°C = −1.36°C = 29.6°F

If the sugar solution is placed outside, where the temperature is 20°F, the solution will freeze.

90. Freezing point depression is 5.4°C

(a) K_b (for H_2O) $= \dfrac{0.512\text{°C kg solvent}}{\text{mol solute}} = \dfrac{0.512\text{°C}}{m}$

$\Delta t_b = mK_b = (2.9 \ m)\left(\dfrac{0.512\text{°C}}{m}\right) = 1.5\text{°C}$

Boiling point = 100.0°C + 1.5°C = 101.5°C

(b) $\Delta t_f = mK_f$

$m = \dfrac{\Delta t_f}{K_f} = \dfrac{5.4 \text{ °C}}{1.86\text{°C kg solvent/mol solute}} = 2.9 \ m$

91. (a) $\left(\dfrac{100.0 \text{ g } C_2H_6O_2}{150.0 \text{ g } H_2O}\right)\left(\dfrac{1 \text{ mol}}{62.07 \text{ g}}\right)\left(\dfrac{1000 \text{ g}}{\text{kg}}\right) = 10.74 \ m$

(b) $\Delta t_b = mK_b = (10.74 \ m)\left(\dfrac{0.512\text{°C}}{m}\right) = 5.50\text{°C}$

Boiling point = 100.00°C + 5.50°C = 105.50°C

(c) $\Delta t_f = mK_f = (10.74 \ m)\left(\dfrac{1.86\text{°C}}{m}\right) = 20.0\text{°C}$

Freezing point = 0.00°C − 20.0°C = −20.0°C

92. (a) $\left(\dfrac{2.68 \text{ g } C_{10}H_8}{38.4 \text{ g } C_6H_6}\right)\left(\dfrac{1 \text{ mol}}{128.2 \text{ g}}\right)\left(\dfrac{1000 \text{ g}}{\text{kg}}\right) = 0.544 \ m$

(b) K_f (for benzene) $= \dfrac{5.1\text{°C}}{m}$ Freezing point of benzene = 5.5°C

$\Delta t_f = (0.544 \ m)\left(\dfrac{5.1\text{°C}}{m}\right) = 2.8\text{°C}$

Freezing point of solution = 5.5°C − 2.8°C = 2.7°C

(c) K_b (for benzene) $= \dfrac{2.53\text{°C}}{m}$ Boiling point of benzene = 80.1°C

$\Delta t_b = (0.544 \ m)\left(\dfrac{2.53\text{°C}}{m}\right) = 1.38\text{°C}$

Boiling point of solution = 80.1°C + 1.38°C = 81.5°C

93. Freezing point of acetic acid is 16.6°C $\quad K_f = \dfrac{3.90°C}{m}$

$\Delta t_f = 16.6°C - 13.2°C = 3.4°C$

$\Delta t_f = mK_f$

$m = \dfrac{3.4°C}{3.90°C /m} = 0.87\ m$

$\left(\dfrac{8.00\ g\ unknown}{60.0\ g\ HC_2H_3O_2}\right)\left(\dfrac{1000\ g}{kg}\right)\left(\dfrac{1\ kg\ HC_2H_3O_2}{0.87\ mol\ unknown}\right) = 153\ g/mol$

94. $\Delta t_f = 2.50°C \quad K_f\ (for\ H_2O) = \dfrac{1.86°C}{m}$

$\Delta t_f = mK_f$

$m = \dfrac{2.50°C}{1.86°C/m} = 1.34\ m$

$\left(\dfrac{4.80\ g\ unknown}{22.0\ g\ H_2O}\right)\left(\dfrac{1000\ g}{kg}\right)\left(\dfrac{1\ kg\ H_2O}{1.34\ mol\ unknown}\right) = 163\ g/mol$

95. Freezing point depression $= 0.372°C \quad K_f = \dfrac{1.86°C}{m}$

$\Delta t_f = mK_f$

$m = \dfrac{0.373°C}{1.86°C/mol} = 0.200\ m$

$(6.20\ g\ C_2H_6O_2)\left(\dfrac{1\ mol}{62.07\ g}\right) = 0.100\ mol\ C_2H_6O_2$

$(0.100\ mol\ C_2H_6O_2)\left(\dfrac{1\ kg\ H_2O}{0.200\ mol\ C_2H_6O_2}\right)\left(\dfrac{1000\ g\ H_2O}{kg\ H_2O}\right) = 500.\ g\ H_2O$

96. (a) Freezing point depression $= 20.0°C$

$12.0\ L\ H_2O = 12,000\ g\ H_2O$

$\Delta t_f = mK_f$

$m = \dfrac{20.0°C}{1.86°C/m} = 10.8\ m$

$(12,000\ g\ H_2O)\left(\dfrac{10.8\ mol\ C_2H_6O_2}{1000\ g\ H_2O}\right)\left(\dfrac{62.07\ g}{mol}\right) = 8.04 \times 10^3\ g\ C_2H_6O_2$

(b) $(8.04 \times 10^3\ g\ C_2H_6O_2)\left(\dfrac{1.00\ mL}{1.11\ g}\right) = 7.24 \times 10^3\ mL\ C_2H_6O_2$

(c) $1.8(-20.0) + 32 = -4.0°F$

97. $NaOH + HCl \longrightarrow NaCl + H_2O$

$$(0.15 \text{ L HCl})\left(\frac{1.0 \text{ mol}}{L}\right)\left(\frac{1 \text{ mol NaOH}}{1 \text{ mol HCl}}\right)\left(\frac{40.0 \text{ g}}{mol}\right) = 6.0 \text{ g NaOH}$$

$$\frac{6.0 \text{ g NaOH}}{x} = \frac{10.0 \text{ g NaOH}}{100.0 \text{ g } 10\% \text{ NaOH solution}}$$

$x = 60. \text{ g } 10\% \text{ NaOH solution}$

98. $NaOH + HCl \longrightarrow NaCl + H_2O$

$$1.0 \ m \ HCl = \frac{1 \text{ mol HCl}}{1 \text{ kg H}_2\text{O}} = \frac{36.46 \text{ g HCl}}{1000 \text{ g H}_2\text{O}}$$

Total mass of solution $= 1000 \text{ g} + 36.46 \text{ g} = 1036.46 \text{ g}$

Therefore, $1.0 \ m \ HCl = \frac{1 \text{mol HCl}}{1036.46 \text{ g HCl solution}}$

$$(250. \text{ g solution})\left(\frac{1 \text{ mol HCl}}{1036.46 \text{ solution}}\right)\left(\frac{1 \text{ mol NaOH}}{1 \text{ mol HCl}}\right)\left(\frac{40.00 \text{ g}}{mol}\right) = 9.65 \text{ g NaOH}$$

$$\frac{9.65 \text{ g NaOH}}{x} = \frac{10. \text{ g NaOH}}{100.0 \text{ g } 10\% \text{ NaOH solution}}$$

$x = 96 \text{ g } 10.\% \text{ NaOH solution}$

99. (a) $(1.0 \text{ L syrup})\left(\frac{1000 \text{ mL}}{L}\right)\left(\frac{1.06 \text{ g}}{mL}\right)\left(\frac{15.0 \text{ g sugar}}{100. \text{ g syrup}}\right) = 1.6 \times 10^2 \text{ g sugar}$

(b) $\left(\frac{1.6 \times 10^2 \text{ g C}_{12}\text{H}_{22}\text{O}_{11}}{L}\right)\left(\frac{1 \text{ mol}}{342.3 \text{ g}}\right) = 0.47 \text{ M}$

(c) $m = \dfrac{\text{mol sugar}}{\text{kg H}_2\text{O}}$

$$\left(\frac{15.0 \text{ g C}_{12}\text{H}_{22}\text{O}_{11}}{85.0 \text{ H}_2\text{O}}\right)\left(\frac{100.0 \text{ g H}_2\text{O}}{1 \text{ kg H}_2\text{O}}\right)\left(\frac{1 \text{ mol}}{342.3 \text{ g}}\right) = 0.516 \text{ m}$$

100. $K_f = \dfrac{5.1°C}{m} \qquad \Delta t_f = 0.614°C$

$$\left(\frac{3.84 \text{ g C}_4\text{H}_2\text{N}}{250. \text{ g C}_6\text{H}_6}\right)\left(\frac{1000 \text{ g}}{kg}\right) = \frac{15.4 \text{ g C}_4\text{H}_2\text{N}}{\text{kg C}_6\text{H}_6}$$

$\Delta t_f = mK_f$

$$m = \frac{0.614°C}{5.1°C/m} = 0.12 \ m = \frac{0.12 \text{ mol C}_4\text{H}_2\text{N}}{\text{kg C}_6\text{H}_6}$$

$$\left(\frac{15.4 \text{ g C}_4\text{H}_2\text{N}}{\text{kg C}_6\text{H}_6}\right)\left(\frac{1 \text{ kg C}_6\text{H}_6}{0.12 \text{ mol C}_4\text{H}_2\text{N}}\right) = 1.3 \times 10^2 \text{ g/mol}$$

Empirical mass $= 64.07 \text{ g}$

$\dfrac{130 \text{ g}}{64.07 \text{ g}} = 2.03$ number of empirical formulas per molecular formula.

Therefore, the molecular formula is twice the empirical formula, or $C_8H_4N_2$.

101. $(12.0 \text{ mol HCl})\left(\frac{36.46 \text{ g}}{\text{mol}}\right) = 438 \text{ g HCl in } 1.00 \text{ L solution}$

$(1.00 \text{ L})\left(\frac{1.18 \text{ g solution}}{\text{mL}}\right)\left(\frac{1000 \text{ mL}}{\text{L}}\right) = 1180 \text{ g solution/L}$

$1180 \text{ g solution} - 438 \text{ g HCl} = 742 \text{ g H}_2\text{O}$

$\text{Since molality} = \frac{\text{mol HCl}}{\text{kg H}_2\text{O}} = \frac{12.0 \text{ mol HCl}}{0.742 \text{ kg H}_2\text{O}} = 16.2 \ m \ \text{HCl}$

102. $\left(\frac{5.5 \text{ mg K}^+}{\text{mL}}\right)\left(\frac{1 \text{ g}}{1000 \text{ mg}}\right)\left(\frac{101.1 \text{ g KNO}_3}{39.10 \text{ g K}^+}\right)(450 \text{ mL}) = 6.4 \text{ g KNO}_3$

$(6.4 \text{ g KNO}_3)\left(\frac{1 \text{ mol}}{101.1 \text{ g}}\right) = 0.063 \text{ mol KNO}_3$

$\frac{0.063 \text{ mol KNO}_3}{0.450 \text{ L}} = 0.14 \text{ M}$

IONIZATION: ACIDS, BASES, SALTS

1. The Arrhenius definition is restricted to aqueous solutions, while the Bronsted-Lowry definition is not.

2. An electrolyte must be present in the solution for the bulb to glow.

3. Electrolytes include acids, bases, and salts.

4. The orientation of the polar water molecules about the Na^+ and Cl^- is different. The positive end (hydrogen) of the water molecule is directed towards Cl^-, while the negative end (oxygen) of the water molecule is directed towards the Na^+. More water molecules will fit around Cl^-, since it is larger than the Na^+ ion (see the illustration on page 102).

5. The pH for a solution with a hydrogen ion concentration of 0.003 M will be between 2 and 3.

6. Tomato juice is more acidic than blood, since its pH is lower.

7. By the Arrhenius theory, an acid is a substance that produces hydrogen ions. A base is a substance that produces hydroxide ions in aqueous solution. By the Bronsted-Lowry theory, an acid is a proton donor, while a base accepts protons. Since a proton is a hydrogen ion, then the two theories are very similar for acids, but not for bases. A chloride ion can accept a proton (producing HCl), so it is a Bronsted-Lowry base, but would not be a base by the Arrhenius theory, since it does not produce hydroxide ions.

    By the Lewis theory, an acid is an electron pair acceptor, and a base is an electron pair donor. Many individual substances would be similarly classified as bases by Bronsted-Lowry or Lewis theories, since a substance with an electron pair to donate, can accept a proton. But, the Lewis definition is almost exclusively applied to reactions where the acid and base combine into a single molecule. The Bronsted-Lowry definition is usually applied to reactions that involve a transfer of a proton from the acid to the base. The Arrhenius definition is most often applied to individual substances, not to reactions. According to the Arrhenius theory, neutralization involves the reaction between a hydrogen ion and a hydroxide ion to form water. Neutralization, according to the Bronsted-Lowry theory, involves the transfer of a proton to a negative ion. The formation of a coordinate-covalent bond constitutes a Lewis neutralization.

8. Neutralization reactions:

    Arrhenius: $HCl + NaOH \longrightarrow NaCl + H_2O$ $(H^+ + OH^- \longrightarrow H_2O)$

    Bronsted-Lowry: $HCl + KCN \longrightarrow HCN + KCl$ $(H^+ + CN^- \longrightarrow HCN)$

    Lewis: $AlCl_3 + NaCl \longrightarrow AlCl_4^- + Na^+$

$$\begin{array}{c} :\ddot{\text{Cl}}: \\ \text{Al}:\ddot{\text{Cl}}: \\ :\ddot{\text{Cl}}: \end{array} \ +\ \left[\ :\ddot{\text{Cl}}:\ \right]^{1-} \longrightarrow \left[\ \begin{array}{c} :\ddot{\text{Cl}}: \\ :\ddot{\text{Cl}}:\text{Al}:\ddot{\text{Cl}}: \\ :\ddot{\text{Cl}}: \end{array}\ \right]^{1-}$$

9. Conjugate acid-base pairs:

(a) HCl - Cl^- ; NH_4^+ - NH_3

(b) HCO_3^- - CO_3^{2-} ; H_2O - OH^-

(c) H_3O^+ - H_2O ; H_2CO_3 - HCO_3^-

(d) $HC_2H_3O_2$ - $C_2H_3O_2^-$; H_3O^+ - H_2O

(e) H_2SO_4 - HSO_4^- ; $H_2C_2H_3O_2^+$ - $HC_2H_3O_2$

(f) step 1: H_2SO_4 - HSO_4^- ; H_3O^+ - H_2O

   step 2: HSO_4^- - SO_4^{2-} ; H_3O^+ - H_2O

(g) $HClO_4$ - ClO_4^- ; H_3O^+ - H_2O

(h) H_3O^+ - H_2O ; CH_3OH - CH_3O^-

10. (a) $\left[\ :\ddot{\text{Br}}:\ \right]^-$ (b) $\left[\ :\ddot{\text{O}}:\text{H}\ \right]^-$ (c) $\left[\ :\text{C}:::\text{N}:\ \right]^-$

These ions are considered to be bases according to the Bronsted-Lowry theory, because they can accept a proton at any of their unshared pairs of electrons. They are considered to be bases according to the Lewis acid-base theory, because they can donate an electron pair.

11. Balancing equations

(a) $Mg\ +\ 2\ HCl\ \longrightarrow\ MgCl_2\ +\ H_2$

(b) $BaO\ +\ 2\ HBr\ \longrightarrow\ BaBr_2\ +\ H_2O$

(c) $2\ Al\ +\ 3\ H_2SO_4\ \longrightarrow\ Al_2(SO_4)_3\ +\ 3\ H_2$

(d) $Na_2CO_3\ +\ 2\ HCl\ \longrightarrow\ 2\ NaCl\ +\ H_2O\ +\ CO_2$

(e) $Fe_2O_3\ +\ 6\ HBr\ \longrightarrow\ 2\ FeBr_3\ +\ 3\ H_2O$

(f) $Ca(OH)_2\ +\ H_2CO_3\ \longrightarrow\ CaCO_3\ +\ 2\ H_2O$

(g) $NaOH\ +\ HBr\ \longrightarrow\ NaBr\ +\ H_2O$

(h) $KOH\ +\ HCl\ \longrightarrow\ KCl\ +\ H_2O$

(i) $Ca(OH)_2\ +\ 2\ HI\ \longrightarrow\ CaI_2\ +\ 2\ H_2O$

(j) $Al(OH)_3\ +\ 3\ HBr\ \longrightarrow\ AlBr_3\ +\ 3\ H_2O$

(k) $Na_2O\ +\ 2\ HClO_4\ \longrightarrow\ 2\ NaClO_4\ +\ H_2O$

(l) $3\ LiOH\ +\ FeCl_3\ \longrightarrow\ Fe(OH)_3\ +\ 3\ LiCl$

(m) $2\ NH_4OH\ +\ FeCl_2\ \longrightarrow\ Fe(OH)_2\ +\ 2\ NH_4Cl$

12. The classes of compounds containing electrolytes are acids, bases, and salts.

13. The following compounds are electrolytes:

(a) HCl – acid

(b) CO_2 – acid (anhydride)

(c) $CaCl_2$ – salt

(g) $NaHCO_3$ – salt

(i) $AgNO_3$ – salt

(j) HCOOH – acid

(k) RbOH – base

(l) K_2CrO_4 – salt

14. Names of the compounds in Table 16.3

H_2SO_4	sulfuric acid	H_2CO_3	carbonic acid
HCl	hydrochloric acid	H_2SO_3	sulfurous acid
$HClO_4$	perchloric acid	$H_2C_2O_4$	oxalic acid
KOH	potassium hydroxide	HClO	hypochlorous acid
$Ba(OH)_2$	barium hydroxide	HF	hydrofluoric acid
HNO_3	nitric acid	$HC_2H_3O_2$	acetic acid
HBr	hydrobromic acid	HNO_2	nitrous acid
NaOH	sodium hydroxide	H_2S	hydrosulfuric acid
$Ca(OH)_2$	calcium hydroxide	H_3BO_3	boric acid
		NH_4OH	ammonium hydroxide

15. Hydrogen chloride dissolved in water conducts an electric current. HCl reacts with polar water molecules to produce H_3O^+ and Cl^- ions, which conduct electric current. Benzene is a nonpolar solvent, so it cannot pull the HCl molecules apart. Since there are no ions in the benzene solution, it does not conduct an electric current.

16. In their crystalline structure, salts exist as positive and negative ions in definite geometric arrangement to each other, held together by the attraction of the opposite charges. When dissolved in water, the salt dissociates as the ions are pulled away from each other by the polar water molecules.

17. Testing the electrical conductivity of the solutions shows that CH_3OH is a nonelectrolyte, while NaOH is an electrolyte. This indicates that the OH group in CH_3OH must be covalently bonded to the CH_3 group.

18. Molten NaCl conducts electricity because the ions are free to move. In the solid state, however, the ions are immobile and do not conduct electricity.

19. Dissociation is the breaking free of already existing ions. Ionization is the formation of ions from molecules. The dissolving of NaCl is a dissociation, since the ions already existed. The dissolving of HCl in water is an ionization process, because ions are formed from molecules.

20. Strong electrolytes are those which are essentially 100% ionized or dissociated in water. Weak electrolytes are those which are only slightly ionized in water.

21. Ions are hydrated in solution because there is an electrical attraction between the charged ions and the polar water molecules.

22. Simplified equations showing the dissociation or ionization of compounds in water:

(a) $Cu(NO_3)_2 \longrightarrow Cu^{2+} + 2\, NO_3^{-}$

(b) $HC_2H_3O_2 \longrightarrow H^+ + C_2H_3O_2^{-}$

(c) $HNO_2 \longrightarrow H^+ + NO_2^{-}$

(d) $LiOH \longrightarrow Li^+ + OH^{-}$

(e) $NH_4Br \longrightarrow NH_4^+ + Br^{-}$

(f) $K_2SO_4 \longrightarrow 2\, K^+ + SO_4^{2-}$

(g) $NaClO_3 \longrightarrow Na^+ + ClO_3^{-}$

(h) $K_3PO_4 \longrightarrow 3\, K^+ + PO_4^{3-}$

23. The main distinction between water solutions of strong and weak electrolytes is the degree of ionization of the electrolyte. A solution of an electrolyte contains many more ions than a solution of a nonelectrolyte. Strong electrolytes are essentially 100% ionized. Weak electrolytes are only slightly ionized in water.

24. (a) In a neutral solution, the concentration of H^+ and OH^- are equal.

(b) In an acid solution, the concentration of H^+ is greater than the concentration of OH^-.

(c) In a basic solution, the concentration of OH^- is greater than the concentration of H^+.

25. The net ionic equation for an acid-base reaction in aqueous solution is $H^+ + OH^- \longrightarrow H_2O$.

26. The HCl molecule is polar and, consequently, is much more soluble in the polar solvent, water, than in the nonpolar solvent, benzene.

27. Pure water is neutral because the concentrations of acid and base ions are equal.

28. The correct statements are a, c, f, i, j, k, l, n, p, r, t, u

(b) The Bronsted-Lowry theory of acids and bases is not restricted to aqueous solutions.

(d) Some substances that are acids according to the Lewis theory will also be acids according to the Bronsted-Lowry theory.

(e) An electron pair acceptor is a Lewis acid.

(g) When an ionic compound dissolves in water, the ions separate. This process is called dissociation.

(h) In autoionization of water, the H_2O and OH^- and the H_3O^+ and H_2O constitute conjugate acid-base pairs.

(m) The terms dissociation and ionization are not synonymous. Dissociation is the separation of an ionic compound into its ions. Ionization is the formation of ions.

(o)   The terms strong acid and strong base refer to solutions in which ionization approaches 100%.   Weak acid and weak base, however, refer to solutions in which ionization is very low.

(q)   Ionic reactions may be represented by net ionic equations.

(s)   It is possible to boil seawater at a higher temperature than that required to boil pure water (both at the same atmospheric pressure).

(v)   The Tyndall effect is observable in colloidal dispersions.

29. The fundamental difference between a colloidal dispersion and a true solution lies in the size of the particles.   In a true solution particles are usually ions or hydrated molecules.   In colloidals the particles are aggregates of ions or molecules.

30. Colloids are prepared by two methods:

1)   dispersion in which larger particles are reduced to colloidal size.   This is usually accomplished by mechanical means.

2)   condensation in which small particles coalesce into larger ones of colloidal size.   This is often accomplished by precipitation in a dilute solution.

31. The Tyndall effect is observed when a narrow beam of light is passed through a colloidal suspension.   The light is reflected from the colloidal particles effectively illuminating the path of the light through the liquid.   In a true solution the light path cannot be seen because the dissolved particles are too small to reflect light.

32. <u>Adsorption</u> refers to the adhesion of particles to a surface while <u>absorption</u> refers to the taking in of one material by another.

33. The particles in a colloid remain dispered because:

1)   they are subjected to constant bombardment by the dispersing phase which keeps the particles in motion and prevents settling out.

2)   since colloid particles have the same kind of charge, they repel each other preventing coalescing to larger particles and settling out.

34. Dialysis is the process of removing dissolved solutes from colloidal dispersion by use of a dialyzing membrane.   The dissolved solutes pass through the membrane leaving the colloidal dispersion behind.   Dialysis is used in artificial kidneys to remove soluble waste products from the blood.

35.

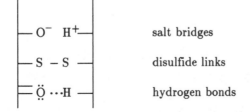

                           salt bridges

                           disulfide links

                           hydrogen bonds

36. Acidic shampoo breaks hydrogen bonds and salt bridges in the hair leaving only disulfide bonds. In a despilatory a basic solution is used which breaks all the types of bonds (H, salt bridges, disulfide) and the hair dissolves.

37. Hair has its maximum strength between pH 4 and pH 5.

38. Calculation of molarity of ions.

(a) $(0.015 \text{ M NaCl})\left(\dfrac{1 \text{ mol Na}^+}{1 \text{ mol NaCl}}\right) = 0.015 \text{ M Na}^+$

$(0.015 \text{ M NaCl})\left(\dfrac{1 \text{ mol Cl}^-}{1 \text{ mol NaCl}}\right) = 0.015 \text{ M Cl}^-$

(b) $(4.25 \text{ M NaKSO}_4)\left(\dfrac{1 \text{ mol Na}^+}{1 \text{ mol NaKSO}_4}\right) = 4.25 \text{ M Na}^+$

$(4.25 \text{ M NaKSO}_4)\left(\dfrac{1 \text{ mol K}^+}{1 \text{ mol NaKSO}_4}\right) = 4.25 \text{ M K}^+$

$(4.25 \text{ M NaKSO}_4)\left(\dfrac{1 \text{ mol SO}_4{}^{2-}}{1 \text{ mol NaKSO}_4}\right) = 4.25 \text{ M SO}_4{}^{2-}$

(c) $(0.75 \text{ M ZnBr}_2)\left(\dfrac{1 \text{ mol Zn}^{2+}}{1 \text{ mol ZnBr}_2}\right) = 0.75 \text{ M Zn}^{2+}$

$(0.75 \text{ M ZnBr}_2)\left(\dfrac{2 \text{ mol Br}^-}{1 \text{ mol ZnBr}_2}\right) = 1.5 \text{ M Br}^-$

(d) $(1.65 \text{ M Al}_2(\text{SO}_4)_3)\left(\dfrac{3 \text{ mol SO}_4{}^{2-}}{1 \text{ mol Al}_2(\text{SO}_4)_3}\right) = 4.95 \text{ M SO}_4{}^{2-}$

$(1.65 \text{ M Al}_2(\text{SO}_4)_3)\left(\dfrac{2 \text{ mol Al}^{3+}}{1 \text{ mol Al}_2(\text{SO}_4)_3}\right) = 3.30 \text{ M Al}^{3+}$

(e) $(0.20 \text{ M CaCl}_2)\left(\dfrac{1 \text{ mol Ca}^{2+}}{1 \text{ mol CaCl}_2}\right) = 0.20 \text{ M Ca}^{2+}$

$(0.20 \text{ M CaCl}_2)\left(\dfrac{2 \text{ mol Cl}^-}{1 \text{ mol CaCl}_2}\right) = 0.40 \text{ M Cl}^-$

(f) $\left(\dfrac{22.0 \text{ g KI}}{500. \text{ mL}}\right)\left(\dfrac{1 \text{ mol}}{166.0 \text{ g}}\right)\left(\dfrac{1000 \text{ mL}}{\text{L}}\right) = 0.265 \text{ M KI}$

$(0.265 \text{ M KI})\left(\dfrac{1 \text{ mol K}^+}{1 \text{ mol KI}}\right) = 0.265 \text{ M K}^+$

$(0.265 \text{ M KI})\left(\dfrac{1 \text{ mol I}^-}{1 \text{ mol KI}}\right) = 0.265 \text{ M I}^-$

(g) $\left(\dfrac{900. \text{ g } (\text{NH}_4)_2\text{SO}_4}{20.0 \text{ L}}\right)\left(\dfrac{1 \text{ mol}}{132.1 \text{ g}}\right) = 0.341 \text{ M } (\text{NH}_4)_2\text{SO}_4$

$(0.341 \text{ M } (\text{NH}_4)_2\text{SO}_4)\left(\dfrac{2 \text{ mol NH}_4{}^+}{1 \text{ mol } (\text{NH}_4)_2\text{SO}_4}\right) = 0.682 \text{ M NH}_4{}^+$

– Chapter 16 –

$$(0.341 \text{ M } (NH_4)_2SO_4)\left(\frac{1 \text{ mol } SO_4^{2-}}{1 \text{ mol } (NH_4)_2SO_4}\right) = 0.341 \text{ M } SO_4^{2-}$$

(h)
$$\left(\frac{0.0120 \text{ g } Mg(ClO_3)_2}{0.00100 \text{ L}}\right)\left(\frac{1 \text{ mol}}{191.2 \text{ g}}\right) = 0.0628 \text{ M } Mg(ClO_3)_2$$
$$(0.0628 \text{ M } Mg(ClO_3)_2)\left(\frac{1 \text{ mol } Mg^{2+}}{1 \text{ mol } Mg(ClO_3)_2}\right) = 0.0628 \text{ M } Mg^{2+}$$

$$(0.0628 \text{ M } Mg(ClO_3)_2)\left(\frac{2 \text{ mol } ClO_3^-}{1 \text{ mol } Mg(ClO_3)_2}\right) = 0.126 \text{ M } ClO_3^-$$

39. The molarities of each ion, as calculated in Exercise 29, will be used to calculate the mass of each ion present in 100. mL of solution.

(a)
$$(0.100 \text{ L})\left(\frac{0.015 \text{ mol } Na^+}{L}\right)\left(\frac{22.99 \text{ g}}{\text{mol}}\right) = 0.034 \text{ g } Na^+$$
$$(0.100 \text{ L})\left(\frac{0.015 \text{ mol } Cl^-}{L}\right)\left(\frac{35.45 \text{ g}}{\text{mol}}\right) = 0.053 \text{ g } Cl^-$$

(b)
$$(0.100 \text{ L})\left(\frac{4.25 \text{ mol } Na^+}{L}\right)\left(\frac{22.99 \text{ g}}{\text{mol}}\right) = 9.77 \text{ g } Na^+$$
$$(0.100 \text{ L})\left(\frac{4.25 \text{ mol } K^+}{L}\right)\left(\frac{39.10 \text{ g}}{\text{mol}}\right) = 16.6 \text{ g } K^+$$
$$(0.100 \text{ L})\left(\frac{4.25 \text{ mol } SO_4^{2-}}{L}\right)\left(\frac{96.06 \text{ g}}{\text{mol}}\right) = 40.8 \text{ g } SO_4^{2-}$$

(c)
$$(0.100 \text{ L})\left(\frac{0.75 \text{ mol } Zn^{2+}}{L}\right)\left(\frac{65.38 \text{ g}}{\text{mol}}\right) = 4.9 \text{ g } Zn^{2+}$$
$$(0.100 \text{ L})\left(\frac{1.5 \text{ mol } Br^-}{L}\right)\left(\frac{79.90 \text{ g}}{\text{mol}}\right) = 12 \text{ g } Br^-$$

(d)
$$(0.100 \text{ L})\left(\frac{3.30 \text{ mol } Al^{3+}}{L}\right)\left(\frac{26.98 \text{ g}}{\text{mol}}\right) = 8.90 \text{ g } Al^{3+}$$
$$(0.100 \text{ L})\left(\frac{4.95 \text{ mol } SO_4^{2-}}{L}\right)\left(\frac{96.06 \text{ g}}{\text{mol}}\right) = 47.5 \text{ g } SO_4^{2-}$$

(e)
$$(0.100 \text{ L})\left(\frac{0.20 \text{ mol } Ca^{2+}}{L}\right)\left(\frac{40.08 \text{ g}}{\text{mol}}\right) = 0.80 \text{ g } Ca^{2+}$$
$$(0.100 \text{ L})\left(\frac{0.400 \text{ mol } Cl^-}{L}\right)\left(\frac{35.45 \text{ g}}{\text{mol}}\right) = 1.4 \text{ g } Cl^-$$

(f)
$$(0.100 \text{ L})\left(\frac{0.265 \text{ mol } K^+}{L}\right)\left(\frac{39.10 \text{ g}}{\text{mol}}\right) = 1.04 \text{ g } K^+$$
$$(0.100 \text{ L})\left(\frac{0.265 \text{ mol } I^-}{L}\right)\left(\frac{126.9 \text{ g}}{\text{mol}}\right) = 3.36 \text{ g } I^-$$

(g)
$$(0.100 \text{ L})\left(\frac{0.682 \text{ mol } NH_4^+}{L}\right)\left(\frac{18.04 \text{ g}}{\text{mol}}\right) = 1.23 \text{ g } NH_4^+$$
$$(0.100 \text{ L})\left(\frac{0.341 \text{ mol } SO_4^{2-}}{L}\right)\left(\frac{96.06 \text{ g}}{\text{mol}}\right) = 3.28 \text{ g } SO_4^{2-}$$

– 127 –

(h) $(0.100 \text{ L})\left(\dfrac{0.0628 \text{ mol Mg}^{2+}}{\text{L}}\right)\left(\dfrac{24.31 \text{ g}}{\text{mol}}\right) = 0.153 \text{ g Mg}^{2+}$

   $(0.100 \text{ L})\left(\dfrac{0.126 \text{ mol ClO}_3^-}{\text{L}}\right)\left(\dfrac{83.45 \text{ g}}{\text{mol}}\right) = 1.05 \text{ g ClO}_3^-$

40. $CaI_2 \longrightarrow Ca^{2+} + 2 I^-$

   $\left(\dfrac{0.520 \text{ mol I}^-}{\text{L}}\right)\left(\dfrac{1 \text{ mol Ca}^{2+}}{2 \text{ mol I}^-}\right) = \left(\dfrac{0.260 \text{ mol Ca}^{2+}}{\text{L}}\right) = 0.260 \text{ M Ca}^{2+}$

41. (a) $(30.0 \text{ mL})\left(\dfrac{1.0 \text{ mol NaCl}}{1000 \text{ mL}}\right) = 0.030 \text{ mol NaCl}$

   $(40.0 \text{ mL})\left(\dfrac{1.0 \text{ mol NaCl}}{1000 \text{ mL}}\right) = 0.040 \text{ mol NaCl}$

   Total mol NaCl $= 0.030 \text{ mol} + 0.040 \text{ mol} = 0.070 \text{ mol NaCl}$

   $\dfrac{0.070 \text{ mol NaCl}}{0.070 \text{ L}} = 1.0 \text{ M NaCl}$

   $(1.0 \text{ M NaCl})\left(\dfrac{1 \text{ mol Na}^+}{1 \text{ mol NaCl}}\right) = 1.0 \text{ M Na}^+$

   $(1.0 \text{ M NaCl})\left(\dfrac{1 \text{ mol Cl}^-}{1 \text{ mol NaCl}}\right) = 1.0 \text{ M Cl}^-$

(b) $HCl + NaOH \longrightarrow NaCl + H_2O$

   $(30.0 \text{ mL HCl})\left(\dfrac{1 \text{ L}}{1000 \text{ mL}}\right)\left(\dfrac{1.0 \text{ mol}}{\text{L}}\right) = 0.030 \text{ mol HCl}$

   $(30.0 \text{ mL NaOH})\left(\dfrac{1 \text{ L}}{1000 \text{ mL}}\right)\left(\dfrac{1.0 \text{ mol}}{\text{L}}\right) = 0.030 \text{ mol NaOH}$

   0.030 mol HCl reacts with 0.030 mol NaOH and produces 0.030 mol NaCl. Final volume, 0.060 L. 0.030 mol NaCl/0.060 L = 0.050 M NaCl. Since there is one mole each of sodium and chloride ions per mole of NaCl, the molar concentrations of Na$^+$ and Cl$^-$ will be 0.050 M Na$^+$, and 0.050 M Cl$^-$.

(c)   100.0 mL of 2.0 M KCl and 100.0 mL of 1.0 M CaCl$_2$ are mixed, giving a final volume of 200.0 mL and concentrations of 1.0 M KCl and 0.5 M CaCl$_2$. The concentration of K$^+$ will be 1.0 M and the concentration of Ca^{2+} will be 0.5 M. The chloride ion concentration will be 2.0 M(1.0 M from the KCl and 2(0.5 M) from the CaCl$_2$).

(d) $KOH + HCl \longrightarrow KCl + H_2O$

   $(100.0 \text{ mL})\left(\dfrac{1 \text{ L}}{1000 \text{ mL}}\right)\left(\dfrac{0.40 \text{ mol KOH}}{\text{L}}\right) = 0.040 \text{ mol KOH}$

   $(100.0 \text{ mL})\left(\dfrac{1 \text{ L}}{1000 \text{ mL}}\right)\left(\dfrac{0.80 \text{ mol HCl}}{\text{L}}\right) = 0.080 \text{ mol HCl}$

   0.040 mol KOH reacts with 0.080 mol HCl. 0.040 mol HCl remains and 0.040 mol KCl is produced. The final volume is 200.0 mL and contains 0.040 mol HCl and 0.040 mol KCl. Moles of ions are: 0.040 mol H$^+$, 0.040 mol K$^+$, and 0.080 mol Cl$^-$. Concentrations of ions are:

   $\dfrac{0.040 \text{ mol H}^+}{0.200 \text{ L}} = 0.20 \text{ M H}^+$ molarity K$^+$ = molarity H$^+$

   $\dfrac{0.080 \text{ mol Cl}^-}{0.200 \text{ L}} = 0.40 \text{ M Cl}^-$

(e) $(35.0 \text{ mL})\left(\dfrac{1 \text{ L}}{1000 \text{ mL}}\right)\left(\dfrac{0.20 \text{ mol Ba(OH)}_2}{\text{L}}\right) = 0.0070 \text{ mol Ba(OH)}_2$

$(35.0 \text{ mL})\left(\dfrac{1 \text{ L}}{1000 \text{ mL}}\right)\left(\dfrac{0.20 \text{ mol H}_2\text{SO}_4}{\text{L}}\right) = 0.0070 \text{ mol H}_2\text{SO}_4$

Final volume = 35.0 mL + 35.0 mL = 70.0 mL

$$\text{H}_2\text{SO}_4 \quad + \quad \text{Ba(OH)}_2 \longrightarrow \text{BaSO}_{4(s)} \quad + \quad 2 \text{ H}_2\text{O}$$

0.0070 mol 　　　　　　 0.0070 mol 　　　 0.0070 mol

The H_2SO_4 and the Ba(OH)_2 react completely producing insoluble BaSO_4 and H_2O. No ions are present in solution.

(f) $(0.500 \text{ L NaCl})\left(\dfrac{2.0 \text{ mol}}{\text{L}}\right) = 1.0 \text{ mol NaCl}$

$(1.00 \text{ L AgNO}_3)\left(\dfrac{1.00 \text{ mol}}{\text{L}}\right) = 1.0 \text{ mol AgNO}_3$

$$\text{NaCl} \quad + \quad \text{AgNO}_3 \longrightarrow \text{AgCl}_{(s)} \quad + \quad \text{NaNO}_3$$

1.0 mol 　　　　　　 1.0 mol 　　　 1.0 mol 　　 1.0 mol

The AgCl is insoluble and produces no ions. The 1.0 mol NaNO_3 will produce 1.0 mol Na^+ ions and 1.0 mol $\text{NO}_3{}^-$ ions. The final volume of the solution is 1.5 L. The concentration of ions are:

$\dfrac{1.0 \text{ mol Na}^+}{1.5 \text{ L}} = 0.67 \text{ M Na}^+$ 　　　　 $\dfrac{1.0 \text{ mol NO}_3{}^-}{1.5 \text{ L}} = 0.67 \text{ M NO}_3{}^-$

42. Dilution problem

$$V_1 M_1 = V_2 M_2$$

$$(100. \text{ mL})(12 \text{ M}) = (V_2)(0.40 \text{ M})$$

$$V_2 = \dfrac{(100. \text{ mL})(12 \text{M})}{0.40 \text{ M}} = 3.0 \times 10^3 \text{ mL}$$

43. The reaction of HCl and NaOH occurs on a 1:1 mole ratio.

$$\text{HCl} + \text{NaOH} \longrightarrow \text{NaCl} + \text{H}_2\text{O}$$

At the endpoint in these titration reactions, equal moles of HCl and NaOH will have reacted. Moles = (molarity)(volume). At the endpoint, mol HCl = mol NaOH. Therefore, at the endpoint, (volume of the acid)(molarity of the acid) = (volume of the base)(molarity of the base). This expression is used in calculating the molarity of the HCl in parts (a), (b), (c), and the molarity of the NaOH in parts (d), (e), (f).

(a) $(37.70 \text{ mL})(0.728 \text{ M}) = (40.13 \text{ mL})(\text{M HCl})$

$\text{M HCl} = \dfrac{(37.70 \text{ mL})(0.728 \text{ M})}{40.13 \text{ mL}} = 0.684 \text{ M HCl}$

(b) $\dfrac{(33.66 \text{ mL})(0.306 \text{ M})}{19.00 \text{ mL}} = 0.542 \text{ M HCl}$

(c) $\dfrac{(18.00 \text{ mL})(0.555 \text{ M})}{27.25 \text{ mL}} = 0.367 \text{ M HCl}$

(d) $\dfrac{(37.19 \text{ mL})(0.126 \text{ M})}{31.91 \text{ mL}} = 0.147 \text{ M NaOH}$

(e) $\dfrac{(48.04 \text{ mL})(0.482 \text{ M})}{24.02 \text{ mL}} = 0.964 \text{ M NaOH}$

(f) $\dfrac{(13.13 \text{ mL})(1.425 \text{ M})}{39.39 \text{ mL}} = 0.4750 \text{ M NaOH}$

44. $Ba(OH)_2 + 2 HCl \longrightarrow BaCl_2 + 2 H_2O$

   M HCl $\longrightarrow$ mol HCl $\longrightarrow$ mol $Ba(OH)_2 \longrightarrow$ M $Ba(OH)_2$

   $\left(\dfrac{0.430 \text{ mol HCl}}{L}\right)\left(\dfrac{1 \text{ L}}{1000 \text{ mL}}\right)(29.26 \text{ mL}) = 0.0126 \text{ mol HCl}$

   $(0.0126 \text{ mol HCl})\left(\dfrac{1 \text{ mol Ba(OH)}_2}{2 \text{ mol HCl}}\right) = 0.00630 \text{ mol Ba(OH)}_2$

   $\dfrac{0.00630 \text{ mol Ba(OH)}_2}{0.02040 \text{ L}} = 0.309 \text{ M Ba(OH)}_2$

45. The acetic acid solution freezes at a lower temperature than the alcohol solution. The acetic acid ionizes slightly, while the alcohol does not. The ionization of the acetic acid increases its particle concentration above that of the alcohol solution, resulting in a lower freezing point for the acetic acid solution.

46. It is more economical to purchase CH_3OH at the same cost per pound as C_2H_5OH. The CH_3OH solution will contain more particles per pound and therefore, have a greater effect on the freezing point of the radiator solution.

47. A hydronium ion is a hydrated hydrogen ion.

   $H^+ \quad + \quad H_2O \quad \longrightarrow \quad H_3O^+$

   (hydrogen ion)   (hydronium ion)

48. Freezing point depression is directly related to the concentration of particles in the solution.

   $C_{12}H_{22}O_{11} \quad > \quad HC_2H_3O_2 > \quad HCl \quad > \quad CaCl_2$

   1 mole $> \quad 1^+$ mol $\quad > \quad$ 2 mol $\quad >$ 3 mol   (particles in solution)

49. (a)   100°C $pH = -\log(1 \times 10^{-6}) = 6$ pH of H_2O is greater at 25°C
           25°C $pH = -\log(1 \times 10^{-7}) = 7$

   (b) $1 \times 10^{-6} > 1 \times 10^{-7}$ so, H^+ concentration is higher at 100°C.

   (c)   The water is neutral at both temperatures, because the H_2O dissociates into equal concentrations of H^+ and OH^-.

50. As the pH changes by 1 unit, the concentration of H^+ in solution changes by a factor of 10. For example, the pH of 0.10 M HCl is 1.0, while the pH of 0.010 M HCl is 2.0.

51. (a) $SO_4^{2-} + Ba^{2+} \longrightarrow BaSO_4(s)$

    (b) $CO_3^{2-} + 2\ H^+ \longrightarrow CO_2(g) + H_2O(l)$

    (c) $Mg(s)\ \ + 2\ HC_2H_3O_2 \longrightarrow Mg^{2+} + H_2(g) + 2\ C_2H_3O_2^-$

    (d) $H_2S(g) + Cd^{2+} \longrightarrow CdS(s) + 2\ H^+$

    (e) $Zn(s) + 2\ H^+ \longrightarrow Zn^{2+} + H_2(g)$

    (f) $Al^{3+} + PO_4^{3-} \longrightarrow AlPO_4(s)$

52. The more acidic solution is listed followed by an explanation.

    (a)   1 molar H_2SO_4. The concentration of H^+ in 1 M H_2SO_4 is greater than 1 M, since there are two ionizable hydrogens per mole of H_2SO_4. In HCl the concentration of H^+ will be 1 M, since there is only one ionizable hydrogen per mole of HCl.

    (b)   1 molar HCl. HCl is a strong electrolyte, while $HC_2H_3O_2$ is a weak electrolyte.

    (c)   2 molar HCl. 2 M HCl will yield 2 M H^+ concentration. 1 M HCl will yield 1 M H^+ concentration.

    (d)   1 molar H_2SO_4. 1 M H_2SO_4 is equivalent to 2 N H_2SO_4 and will therefore have a greater concentration of H^+ than 1 N H_2SO_4.

53. (a) $2\ HCl + Ca(OH)_2 \longrightarrow CaCl_2 + 2\ H_2O$

          $M\ Ca(OH)_2 \longrightarrow mol\ Ca(OH)_2 \longrightarrow mol\ HCl \longrightarrow mL\ HCl$

$$(0.0500\ L\ Ca(OH)_2)\left(\frac{0.100\ mol}{L}\right)\left(\frac{2\ mol\ HCl}{1\ mol\ Ca(OH)_2}\right)\left(\frac{1000\ mL}{0.245\ mol}\right) = 40.8\ mL\ of\ 0.245\ M\ HCl$$

    (b) $3\ HCl + Al(OH)_3 \longrightarrow AlCl_3 + 3\ H_2O$

          $g\ Al(OH)_3 \longrightarrow mol\ Al(OH)_3 \longrightarrow mol\ HCl \longrightarrow mL\ HCl$

$$(10.0\ g\ Al(OH)_3)\left(\frac{1\ mol}{78.01\ g}\right)\left(\frac{3\ mol\ HCl}{1\ mol\ Al(OH)_3}\right)\left(\frac{1000\ ml}{0.245\ mol}\right) = 1.57\ \times\ 10^3\ mL\ of\ 0.245\ M\ HCl$$

54. $Na_2CO_3 + 2\ HCl \longrightarrow 2\ NaCl + CO_2 + H_2O$

     $g\ Na_2CO_3 \longrightarrow mol\ Na_2CO_3 \longrightarrow mol\ HCl \longrightarrow M\ HCl$

$$(0.452\ g\ Na_2CO_3)\left(\frac{1\ mol}{106.0\ g}\right)\left(\frac{2\ mol\ HCl}{1\ mol\ Na_2CO_3}\right) = 0.00853\ mol\ HCl$$

$$\frac{0.00853\ mol}{0.0424\ L} = 0.201\ M\ HCl$$

55. $2\ HCl + Ca(OH)_2 \longrightarrow CaCl_2 + 2\ H_2O$

     $g\ Ca(OH)_2 \longrightarrow mol\ Ca(OH)_2 \longrightarrow mol\ HCl \longrightarrow mL\ HCl$

$$(2.00\ g\ Ca(OH)_2)\left(\frac{1\ mol}{74.10\ g}\right)\left(\frac{2\ mol\ HCl}{1\ mol\ Ca(OH)_2}\right)\left(\frac{1000\ mL}{0.1234\ mol}\right) = 437\ mL\ of\ 0.1234\ M\ HCl$$

56. $KOH + HNO_3 \longrightarrow KNO_3 + H_2O$

L $HNO_3 \longrightarrow$ mol $HNO_3 \longrightarrow$ mol $KOH \longrightarrow$ g KOH

$(0.05000 \text{ L } HNO_3)\left(\dfrac{0.240 \text{ mol}}{L}\right)\left(\dfrac{1 \text{ mol } KOH}{1 \text{ mol } HNO_3}\right)\left(\dfrac{56.11 \text{ g}}{\text{mol}}\right) = 0.673 \text{ g } KOH$

57. $NaOH + HCl \longrightarrow NaCl + H_2O$

L $HCl \longrightarrow$ mol $HCl \longrightarrow$ mol $NaOH \longrightarrow$ g $NaOH$

$(0.01825 \text{ L } HCl)\left(\dfrac{0.2406 \text{ mol}}{L}\right)\left(\dfrac{1 \text{ mol } NaOH \text{ mol}}{1 \text{ mol } HCl}\right)\left(\dfrac{40.00 \text{ g}}{\text{mol}}\right) = 0.176 \text{ g } NaOH$

$\left(\dfrac{0.176 \text{ g } NaOH}{0.200 \text{ g sample}}\right)(100) = 88.0\% \text{ NaOH}$

58. $NaOH + HCl \longrightarrow NaCl + H_2O$

L $HCl \longrightarrow$ mol $HCl \longrightarrow$ mol $NaOH \longrightarrow$ g $NaOH$

$(0.04990 \text{ L } HCl)\left(\dfrac{0.466 \text{ mol}}{L}\right)\left(\dfrac{1 \text{ mol } NaOH}{1 \text{ mol } HCl}\right)\left(\dfrac{40.00 \text{ g}}{\text{mol}}\right) = 0.930 \text{ g } NaOH$

1.00 g sample $-$ 0.930 g $NaOH$ = 0.070 g $NaCl$ in sample

$\left(\dfrac{0.070 \text{ g } NaCl}{1.00 \text{ g sample}}\right)(100) = 7.0\% \text{ NaCl in sample}$

59. $Zn + 2 HCl \longrightarrow ZnCl_2 + H_2$

g $Zn \longrightarrow$ mol Zn

$(5.00 \text{ g } Zn)\left(\dfrac{1 \text{ mol}}{65.38 \text{ g}}\right) = 0.0765 \text{ mol } Zn$

(a) $(0.100 \text{ L } HCl)\left(\dfrac{0.350 \text{ mol}}{L}\right) = 0.0350 \text{ mol } HCl$

The HCl is the limiting reactant.

$(0.0350 \text{ mol } HCl)\left(\dfrac{1 \text{ mol } H_2}{2 \text{ mol } HCl}\right) = 0.0175 \text{ mol } H_2$

$T = 27°C = 300. \text{ K}$

$P = (700. \text{ torr})\left(\dfrac{1 \text{ atm}}{760 \text{ torr}}\right) = 0.921 \text{ atm}$

$PV = nRT$

$V = \dfrac{nRT}{P} = \dfrac{(0.0175 \text{ mol})(0.0821 \text{ L atm/mol K})(300. \text{ K})}{0.921 \text{ atm}} = 0.468 \text{ L } H_2$

(b) $(0.200 \text{ L } HCl)\left(\dfrac{0.350 \text{ mol}}{L}\right) = 0.0700 \text{ mol } HCl$

The HCl is the limiting reactant.

$(0.0700 \text{ mol } HCl)\left(\dfrac{1 \text{ mol } H_2}{2 \text{ mol } HCl}\right) = 0.0350 \text{ mol } H_2$

$$T = 27°C = 300.\ K$$
$$P = 0.921\ \text{atm (from part a)}$$
$$V = \frac{nRT}{P} = \frac{(0.0350\ \text{mol})(0.0821\ \text{L atm/mol K})(300.\ K)}{0.921\ \text{atm}} = 0.936\ \text{L}\ H_2$$

60. Calculation of the pH solutions:

(a) $H^+ = 0.01\ M = 10^{-2}\ M;\ pH = -\log 10^{-2} = 2$

(b) $H^+ = 1.0\ M;\ pH = -\log 1.0 = 0$

(c) $H^+ = 6.5\ \text{x}\ 10^{-9}\ M;\ pH = -\log (6.5\ \text{x}\ 10^{-9}) = 8.19$

(d) $H^+ = 1\ \text{x}\ 10^{-7}\ M;\ pH = -\log (1\ \text{x}\ 10^{-7}) = 7$

(e) $H^+ = 0.50\ M;\ pH = -\log (5.0\ \text{x}\ 10^{-1}) = 0.30$

(f) $H^+ = 0.00010\ M = 1.0\ \text{x}\ 10^{-4}\ M;\ pH = -\log (1.0\ \text{x}\ 10^{-4}) = 4.0$

61. (a) Orange juice $= 3.7\ \text{x}\ 10^{-4}\ M\ H^+$

$pH = -\log (3.7\ \text{x}\ 10^{-4}) = 3.4$

(b) Vinegar $= 2.8\ \text{x}\ 10^{-3}\ M\ H^+$

$pH = -\log (2.8\ \text{x}\ 10^{-3}) = 2.6$

(c) Black coffee $= 5.0\ \text{x}\ 10^{-5}\ M\ H^+$

$pH = -\log (5.0\ \text{x}\ 10^{-5}) = 4.3$

(d) Limewater $= 3.4\ \text{x}\ 10^{-11}\ M\ H^+$

$pH = -\log (3.4\ \text{x}\ 10^{-11}) = 10.5$

62. pH of 1.0 L solution containing 0.1 mL of 1.0 M HCl

$$(0.1\ \text{mL})\left(\frac{1\ L}{1000\ \text{mL}}\right)\left(\frac{1.0\ \text{mol HCl}}{L}\right) = 1\ \text{x}\ 10^{-4}\ M\ HCl$$

$1.0\ \text{x}\ 10^{-4}\ M\ HCl$ produces $1\ \text{x}\ 10^{-4}\ M\ H^+$

$pH = -\log (1\ \text{x}\ 10^{-4}) = 4$

63. $V_1M_1 = V_2M_2 \quad V_1 = \frac{V_2M_2}{M_1} = \frac{(50.0\ L)(5.00\ M)}{18.0\ M} = 13.9\ \text{L of 18.0 M}\ H_2SO_4$

64. $NaOH + HCl \longrightarrow NaCl + H_2O$

$$(3.0\ \text{g NaOH})\left(\frac{1\ \text{mol}}{40.00\ \text{g}}\right) = 0.075\ \text{mol NaOH}$$

$$(500\ \text{mL HCl})\left(\frac{1\ L}{1000\ \text{mL}}\right)\left(\frac{0.10\ \text{mol}}{L}\right) = 0.050\ \text{mol HCl}$$

The solution is basic. The NaOH will neutralize the HCl with an excess of 0.025 moles of NaOH remaining.

65. $V_AN_A = V_BN_B \quad V_B = \frac{V_AN_A}{V_B} = \frac{(28.92\ \text{mL})(0.1240\ N)}{10.00\ \text{mL}} = 0.3586\ N\ NaOH$

66. $V_AN_A = V_BN_B \quad V_A = \frac{V_BN_B}{N_A} = \frac{(32.8\ \text{mL})(0.225\ N)}{0.325\ N} = 22.7\ \text{mL of 0.325 N}\ HNO_3$

67. $V_A N_A = V_B N_B$ $\quad V_A = \dfrac{V_B N_B}{N_A} = \dfrac{(32.8\ \text{mL})(0.225\ \text{N})}{0.325\ \text{N}} = 22.7\ \text{mL of } 0.325\ \text{N}\ H_2SO_4$

68. $V_A N_A = V_B N_B$ $\quad N_A = \dfrac{V_B N_B}{V_A} = \dfrac{(22.68\ \text{mL})(0.5000\ \text{N})}{25.00\ \text{mL}} = 0.4536\ \text{N}\ H_3PO_4$

$$\left(\dfrac{0.4536\ \text{eq}\ H_3PO_4}{L}\right)\left(\dfrac{1\ \text{mol}}{3\ \text{eq}}\right) = 0.1512\ \text{M}\ H_3PO_4$$

69. $2\ NaOH + H_2SO_4 \longrightarrow H_2O + 2\ Na_2SO_4$

$\quad N_A V_A = N_B V_B$

$\quad (0.20\ N_A)(60.\ \text{mL}) = (0.10\ N_B(V_B)$

$\quad 1.2 \times 10^2\ \text{mL} = V_B$

70. (a) $\quad N_A V_A = N_B V_B$

$\quad N_A(25\ \text{mL}) = (0.20\ N_B)(40.\ \text{mL})$

$\quad N_A = 0.32\ \text{N}\ H_2SO_4$

(b) $\quad \left(\dfrac{0.32\ \text{eq}}{L}\right)\left(\dfrac{1\ \text{mol}}{2\ \text{eq}}\right)(0.025\ \text{L})(98.08\ \text{g/mol}) = 0.39\ \text{g}\ H_2SO_4$

71. $HCl + NaOH \longrightarrow NaCl + H_2O$

The residue is NaCl.

g NaCl $\longrightarrow$ mol NaCl $\longrightarrow$ mol HCl or mol NaOH

$\quad (0.117\ \text{g NaCl})\left(\dfrac{1\ \text{mol}}{58.44\ \text{g}}\right)\left(\dfrac{1\ \text{mol HCl}}{1\ \text{mol NaCl}}\right) = 0.00200\ \text{mol HCl or } 0.00200\ \text{mol NaOH}$

Molarity and normality are the same for HCl and NaOH, since each has 1 eq/ mol.

$\dfrac{0.00200\ \text{eq HCl}}{0.0400\ \text{L}} = 0.0500\ \text{N}\ HCl \qquad \dfrac{0.00200\ \text{eq NaOH}}{0.0200\ \text{L}} = 0.100\ \text{N}\ NaOH$

72. Equivalents of acid $\quad = $ equivalents of base

$\qquad\qquad\qquad\quad = N_B V_B$

$\qquad\qquad\qquad\quad = (0.10\ \text{eq/L})(0.025\ \text{L})$

$\qquad\qquad\qquad\quad = 2.5 \times 10^{-3}\ \text{eq}$

Equivalent mass $\quad = \dfrac{0.305\ \text{g}}{2.5 \times 10^{-3}\ \text{eq}} = 1.2 \times 10^2\ \text{g/eq}$

73. Equivalents of acid $\quad = $ equivalents of base

$\qquad\qquad\qquad\quad = N_B V_B$

$\qquad\qquad\qquad\quad = (0.10\ \text{eq/L})(0.125\ \text{L})$

$\qquad\qquad\qquad\quad = 1.3 \times 10^{-2}\ \text{eq}$

Equivalent mass $\quad = \dfrac{0.738\ \text{g}}{1.3 \times 10^{-2}\ \text{eq}} = 57\ \text{g/eq}$

CHEMICAL EQUILIBRIUM

1. Both tubes would appear the same and contain more molecules in the gaseous state than the tube at 0°C, and less molecules in the gaseous state than the tube at 80°C.

2. The reaction is endothermic because the increased temperature increases the concentration of product present at equilibrium.

3. At equilibrium, the rate of the forward reaction equals the rate of the reverse reaction.

4. The yield of NH_3 would be greater if the reaction were carried out in a one liter vessel. The pressure would be greater in the 1 L vessel. Increased pressure favors more product because the reaction of four moles of reactants (3 moles H_2 and 1 mole N_2) produces 2 moles of product, NH_3.

5. Acids stronger than acetic acid are: benzoic, cyanic, formic, hydrofluoric, and nitrous acids (all equilibrium constants are greater than the equilibrium constant for acetic acid). Acids weaker than acetic acid are: carbonic, hydrocyanic, and hypochlorous acids (all have equilibrium constants smaller than the equilibrium constant for acetic acid). All have one ionizable hydrogen atom.

6. The sum of the pH and the pOH is 14. A solution whose pH is -1 would have a pOH of 15.

7. The order of solubility will correspond to the order of the values of the solubility product constants of the salts being compared. This occurs because each salt in the comparison produces the same number of ions (two in this case) for each formula unit of salt that dissolves. This type of comparison would not necessarily be valid if the salts being compared gave different numbers of ions per formula unit of salt dissolving. The order is: $AgC_2H_3O_2$, $PbSO_4$, $AgCl$, $BaCrO_4$, $AgBr$, AgI, PbS.

8. (a) K_{sp} $Mn(OH)_2$ = 2.0 x 10^{-13}; K_{sp} Ag_2CrO_4 = 1.9 x 10^{-12}.

   Each salt gives 3 ions per formula units of salt dissolving. Therefore, the salt with the largest K_{sp} (in this case Ag_2CrO_4) is more soluble.

   (b) K_{sp} $BaCrO_4$ = 8.5 x 10^{-11}; K_{sp} Ag_2CrO_4 = 1.9 x 10^{-12}. Ag_2CrO_4 has a greater molar solubility than $BaCrO_4$, even though its K_{sp} is smaller, because the Ag_2CrO_4 produces more ions per formula unit of salt dissolving than $BaCrO_4$.

   $BaCrO_4(s) \rightleftharpoons Ba^{2+} + CrO_4^{2-}$

   Molar solubility = $\sqrt{8.5 \times 10^{-11}}$ = 9.2 x 10^{-6} mol $BaCrO_4$/L

   $Ag_2CrO_4(s) \rightleftharpoons 2 Ag^+ + CrO_4^{2-}$

   Molar solubility = $\sqrt[3]{\frac{1.9 \times 10^{-12}}{4}}$ = 7.8 x 10^{-5} mol Ag_2CrO_4/L

9. $HC_2H_3O_2 \rightleftharpoons H^+ + C_2H_3O_2^-$

Inital Concentrations		Added	Concentration After Equilibrium Shifts
$HC_2H_3O_2$	1.00 M	------	1.01 M
H^+	1.8×10^{-5} M	0.010 M	1.9×10^{-5}
$C_2H_3O_2^-$	1.00 M	------	0.99 M

The initial concentration of H^+ in the buffer solution is very low (1.8×10^{-5} M) because of the large excess of acetate ions.   0.010 mole of HCl is added to one liter of the buffer solution. This will supply 0.010 M H^+.   The added H^+ creates a stress to the right side of the equation. The equilibrium shifts to the left, using up almost all the added H^+, reducing the acetate ion by approximately 0.010 M, and increasing the acetic acid by approximately 0.010 M.   The concentration of H^+, which will be in equilibrium with these concentrations, calculates to be 1.9×10^{-5} M, which can be verified by experiment.

10. If the reaction shown in Figure 17.4 were exothermic, the figure should be modified to show the energy level of the product as lower than the energy level of the reactants.

11. Reversible systems

   (a) $H_2O(s) \overset{0°C}{\rightleftharpoons} H_2O(l)$

   (b) $H_2O(l) \overset{100°C}{\rightleftharpoons} H_2O(g)$

   (c) $Na_2SO_4(s) \rightleftharpoons 2\ Na^+(aq) + SO_4^{2-}(aq)$

   (d) $SO_2(l) \rightleftharpoons SO_2(g)$

12. If a sodium chloride solution is saturated, the equilibrium is

   $$Na^+(aq) + Cl^-(aq) \rightleftharpoons NaCl(s)$$

   Bubbling in HCl gas increases the concentration of Cl^-, creating a stress, which will cause the equilibrium to shift to the right, precipitating solid NaCl.

13. The rate of a reaction increases when the concentration of one of the reactants increases.   The increase in concentration causes the number of collisions between the reactants to increase.

14. Equilibrium system

   $$4\ NH_3(g) + 3\ O_2(g) \rightleftharpoons 2\ N_2(g) + 6\ H_2O(g) + 1531\ kJ$$

   (a)   The reaction is exothermic with heat being evolved.

   (b)   The addition of O_2 will shift the reaction to the right until equilibrium is reestablished. The concentration of N_2, H_2O, and O_2 will be increased.   The concentration of the NH_3 will be decreased.

   (c)   The addition of heat will shift the reaction to the left.   This will use up the heat added.

15. $N_2(g) + 3 H_2(g) \rightleftharpoons 2 NH_3(g) + 92.5$ kJ

Change or stress imposed on the system at equilibrium	Direction of reaction, left or right, to re-establish equilibrium	Change in number of moles		
		N_2	H_2	NH_3
(a) Add N_2	right	I	D	I
(b) Remove H_2	left	I	D	D
(c) Decrease volume of reaction vessel	right	D	D	I
(d) Increase volume of reaction vessel	left	I	I	D
(e) Increase temperature	left	I	I	D
(f) Add a catalyst	no change	N	N	N
(g) Add H_2 and NH_3	?	?	I	I

I = Increase; D = Decrease; N = No Change;
? = Not sufficient information to determine

16. If pure HI is placed in a vessel at 700 K, some of it will decompose. The mechanism would be for two molecules to collide, forming the activated complex, which would, in turn, decompose to form H_2 and I_2.

17. Direction of shift in equilibrium:

Reaction	Increase Temperature	Increased Pressure (Volume Decrease)	Add Catalyst
(a)	right	right	no change
(b)	left	no change	no change
(c)	left	right	no change
(d)	right	left	no change
(e)	left	left	no change

18. An increase in temperature causes the rate of reaction to increase, because it speeds up the motion of the molecules. Faster moving molecules increase the number and effectiveness of the collisions, resulting in enough energy transfer to cause a reaction.

19. $A + B \rightleftharpoons C + D$

When A and B are initially mixed, the rate of the forward reaction to produce C and D is at its maximum. As the reaction proceeds, the rate of production of C and D decreases because the concentrations of A and B decrease. As C and D are produced, some of the collisions between C and D will result in the reverse reactions, forming A and B. Finally, an equilibrium is achieved in which the forward rate exactly equals the reverse rate.

20. (a) right
  (b) left
  (c) none
  (d) left
  (e) right

21. $HC_2H_3O_2 + H_2O \rightleftharpoons H_3O^+ + C_2H_3O_2^-$

As water is added (diluting the solution from 1.0 M to 0.10 M), the equilibrium shifts to the right, yielding a higher percent ionization.

22. The statement does not contradict Le Chatelier's Principle. The previous question deals with the case of dilution. If pure acetic acid is added to a dilute solution, the reaction will shift to the right, producing more ions in accordance with Le Chatelier's Principle. But, the concentration of the un-ionized acetic acid will increase faster than the concentration of the ions, thus yielding a smaller percent ionization.

23. (a) $K_{eq} = \dfrac{[Cl_2]^2 [H_2O]^2}{[HCl]^4 [O_2]}$

 (d) $K_{eq} = \dfrac{[H^+] [ClO_2^-]}{[HClO_2]}$

 (b) $K_{eq} = \dfrac{[NH_3]^2}{[N_2] [H_2]^3}$

 (e) $K_{eq} = \dfrac{[NH_4^+] [OH^-]}{[NH_4OH]}$

 (c) $K_{eq} = \dfrac{[PCl_3] [Cl_2]}{[PCl_5]}$

 (f) $K_{eq} = \dfrac{[NO]^4 [H_2O]^6}{[NH_3]^4 [O_2]^5}$

24. At different temperatures, the degree of ionization of water varies, being higher at higher temperatures. Consequently, the pH of water can be different at different temperatures.

25. In pure water, H^+ and OH^- are produced in equal quantities by the ionization of the water molecules, $H_2O \rightleftharpoons H^+ + OH^-$. Since pH $= -\log H^+$, and pOH $= -\log OH^-$, they will always be identical for pure water. At 25°C, they each have the value of 7, but at higher temperatures, the degree of ionization is greater, so the pH and pOH would both be less than 7, but still equal.

26. If the H^+ ion concentration is increased:

 (a) pH is decreased (pH of 1 is more acidic than that of 4)
 (b) pOH is increased
 (c) OH^- is decreased
 (d) K_w remains unchanged. It is a constant at a given temperature.

27. (a) $K_{sp} = [Cu^{2+}] [S^{2-}]$

 (e) $K_{sp} = [Fe^{3+}] [OH^-]^3$

 (b) $K_{sp} = [Ba^{2+}] [SO_4^{2-}]$

 (f) $K_{sp} = [Sb^{5+}]^2 [S^{2-}]^5$

 (c) $K_{sp} = [Pb^{2+}] [Br^-]^2$

 (g) $K_{sp} = [Ca^{2+}] [F^-]^2$

 (d) $K_{sp} = [Ag^+]^3 [AsO_4^{3-}]$

 (h) $K_{sp} = [Ba^{2+}]^3 [PO_4^{3-}]^2$

28. In water, the silver acetate dissociates until the equilibrium concentration of ions is reached. In nitric acid solution, the acetate ions will react with hydrogen ions to form acetic acid molecules. Since the acetate ion concentration is kept low, more silver acetate can dissolve. If HCl is used, a precipitate of silver chloride would be formed, since silver chloride is less soluble than silver acetate. Thus, more silver acetate would dissolve in HCl than in pure water.

$$Ag\ C_2H_3O_{2(s)} \rightleftarrows Ag^+ + C_2H_3O_2^-$$

29. The basis for deciding if a salt dissolved in water produces an acidic, a basic, or a neutral solution, is whether or not the salt reacts with water (hydrolysis). Salts that contain an ion derived from a weak acid or base will hydrolyze to produce an acidic or a basic solution.

(a)	KCl, neutral	(e)	$Ca(CN)_2$, basic
(b)	Na_2CO_3, basic	(f)	$BaBr_2$, neutral
(c)	K_2SO_4, neutral	(g)	$NaNO_2$, basic
(d)	$(NH_4)_2SO_4$, acidic	(h)	NaF, basic

30. When the salt, sodium acetate, is dissolved in water, the solution becomes basic. The dissolving reaction is

$$NaC_2H_3O_{2(s)} \xrightarrow{H_2O} Na^+ + C_2H_3O_2^-$$

The acetate ion reacts with water (hydrolysis). The reaction does not go to completion, but some OH^- ions are produced and the solution becomes basic.

$$C_2H_3O_2^- + H_2O \rightleftarrows OH^- + HC_2H_3O_2$$

31. (a) $NO_2^- + H_2O \rightleftarrows OH^- + HNO_2$

    (b) $C_2H_3O_2^- + H_2O \rightleftarrows OH^- + HC_2H_3O_2$

    (c) $NH_4^+ + H_2O \rightleftarrows H^+ + NH_4OH$

    (d) $SO_3^{2-} + 2\ H_2O \rightleftarrows 2\ OH^- + H_2SO_3$

32. (a) $HCO_3^- + H_2O \rightleftarrows OH^- + H_2CO_3$

    (b) $NH_4^+ + H_2O \rightleftarrows H^+ + NH_4OH$

    (c) $OCl^- + H_2O \rightleftarrows OH^- + HOCl$

    (d) $ClO_2^- + H_2O \rightleftarrows OH^- + HClO_2$

33. A buffer solution contains a weak acid or base plus a salt of that weak acid or base, such as dilute acetic acid and sodium acetate.

$$HC_2H_3O_{2(aq)} \rightleftarrows H^+ + C_2H_3O_2^-$$
$$NaC_2H_3O_{2(aq)} \rightleftarrows Na^+ + C_2H_3O_2^-$$

When a small amount of a strong acid (H^+) is added to this buffer solution, the H^+ reacts with the acetate ions to form un-ionized acetic acid, thus neutralizing the added acid. When a strong base, OH^-, is added it reacts with un-ionized acetic acid to neutralize the added base. As a result, in both cases, the approximate pH of the solution is maintained.

34. (a) When excess acid gets into the blood stream, bicarbonate ions react with it to form H_2CO_3, with almost no net increase in H^+, thus maintaining the pH.

(b) When excess base gets into the blood stream, it reacts with H^+ to form water, but by the first reaction, more H_2CO_3 will ionize to replace the H^+ ions, maintaining the pH at a constant value.

35. The equilibrium reaction is

$$Hb + O_2 \rightleftharpoons HbO_2$$

In the lungs the concentration of O_2 is high and favors binding O_2 to hemoglobin. From there it is transported to the tissues.

36. Both molecules bind oxygen. The myoglobin molecule binds oxygen differently than hemoglobin. Since the affintiy between oxygen and myoglobin is higher than that between oxygen and hemoglobin, the hemoglobin will release oxygen to myoglobin for storage.

37. Hemoglobin depleted of oxygen has the potential to carry CO_2. the CO_2 binds at one end of the protein chain at a different site than O_2. When CO_2 dissolves in water the following reaction occurs:

$$CO_2 + H_2O \rightleftharpoons HCO_3^- + H^+$$

To facilitate removal of CO_2 the hemoglobin binds H^+ shifting the equilibrium towards the right.

38. The correct statements are c, d, g, i, j, k, l, n, p, q, s, t, u, v, x, z

(a) In a reaction, at equilibrium, the concetration of reactants and products usually are not equal.

(b) A catalyst has no effect on the concentration of the products at equilibrium.

(e) If an increase in temperature causes an increase in the concentration of products present at equilibrium, the reaction is endothermic.

(f) The magnitude of an equilibrium constant is dependent upon temperature.

(h) The amount of product obtained at equilibrium is independent of the time required to reach equilibrium.

(m) The reaction shown is exothermic.

(o) Increasing the temperature will decrease the magnitude of the equilibrium constant, K_{eq}.

(r) High pressures lead to decreased yields of NO_2.

(w) Addition of solid $BaSO_4$ to a saturated solution of $BaSO_4$ has no effect on the magnitude of K_{sp}.

(y) The pH decreases as the $[H^+]$ increases.

39. $H_2 + I_2 \rightleftharpoons 2\,HI$

The reaction is on a 1 to 1 mole ratio of hydrogen to iodine. The hydrogen is the limiting reactant.

$$(2.10 \text{ mol } H_2)\left(\frac{2 \text{ mol HI}}{1 \text{ mol } H_2}\right) = 4.20 \text{ mol HI}$$

40. $H_2 + I_2 \rightleftharpoons 2 \text{ HI}$

(a) $(2.00 \text{ mol } H_2)\left(\frac{2 \text{ mol HI}}{1 \text{ mol } H_2}\right)\left(\frac{0.79 \text{ mol}}{1.00 \text{ mol}}\right) = 3.16 \text{ mol HI}$

(b) The addition of 0.27 mol I_2 makes the iodine present in excess and the 2.00 mol H_2 the limiting reactant.   The yield increases to 85%.

$$(2.00 \text{ mol } H_2)\left(\frac{2 \text{ mol HI}}{1 \text{ mol } H_2}\right)\left(\frac{0.85 \text{ mol}}{1.00 \text{ mol}}\right) = 3.4 \text{ mol HI}$$

There will be 15% unreacted H_2 and I_2.

$(0.15)(2.0 \text{ mol } H_2) = 0.30 \text{ mol } H_2$ present.

In addition to the 0.30 mol of unreacted I_2, will be the 0.27 mol I_2 added.

$0.27 \text{ mol } + 0.30 \text{ mol } = 0.57 \text{ mol } I_2$ present.

(c) $K = \dfrac{[\text{HI}]^2}{[\text{H}_2]\,[\text{I}_2]}$

The formation of 3.16 mol HI required the reaction of 1.58 mol I_2 and 1.58 mol H_2. At equilibrium, the concentrations are:

3.16 mol HI; $2.00 - 1.58 = 0.42 \text{ mol } H_2 = 0.42 \text{ mol } I_2$

$$K_{eq} = \frac{(3.16)^2}{(0.42)(0.42)} = 57$$

In the calculation of the equilibrium constant, the actual number of moles of reactants and products present at equilibrium can be used in the calculation in place of molar concentrations. This occurs because the reaction is gaseous and the liters of HI produced equals the sum of the liters of H_2 and I_2 reacting.   In the equilibrium expression, the volumes will cancel.

41. $H_2 + I_2 \rightleftharpoons 2 \text{ HI}$

$$(64.0 \text{ g HI})\left(\frac{1 \text{ mol}}{127.9 \text{ g}}\right) = 0.500 \text{ mol HI present}$$

$$(0.500 \text{ mol HI})\left(\frac{1 \text{ mol } I_2}{2 \text{ mol HI}}\right) = 0.250 \text{ mol } I_2 \text{ reacted}$$

$$(0.500 \text{ mol HI})\left(\frac{1 \text{ mol } H_2}{2 \text{ mol HI}}\right) = 0.250 \text{ mol } H_2 \text{ reacted}$$

$$(6.00 \text{ g } H_2)\left(\frac{1 \text{ mol}}{2.16 \text{ g}}\right) = 2.98 \text{ mol } H_2 \text{ initially present}$$

$$(200. \text{ g } I_2)\left(\frac{1 \text{ mol}}{253.8 \text{ g}}\right) = 0.788 \text{ mol } I_2 \text{ initially present}$$

At equilibrium, moles present are:

0.500 mol HI; $2.98 - 0.250 = 2.73 \text{ mol } H_2$

$0.788 - 0.250 = 0.538 \text{ mol } I_2$

42. $PCl_3(g) + Cl_2(g) \rightleftharpoons PCl_5(g)$

$$K_{eq} = \frac{[PCl_5]}{[PCl_3][Cl_2]}$$

The concentrations are:

$$PCl_5 = \frac{0.22 \text{ mol}}{20. \text{ L}} = 0.011 \text{ M}$$

$$PCl_3 = \frac{0.10 \text{ mol}}{20. \text{ L}} = 0.0050 \text{ M}$$

$$Cl_2 = \frac{1.50 \text{ mol}}{20. \text{ L}} = 0.075 \text{ M}$$

$$K_{eq} = \frac{0.011}{(0.0050)(0.075)} = 29$$

43. $100°C - 30°C = 70°C$ temperature increase.   This increase is equal to seven $10°C$ increments. The reaction rate will be increased $2^7 = 128$ times.

44. Ionization constants.

Hypochlorous acid $HOCl \rightleftharpoons H^+ + OCl^-$

Equilibrium concentrations:

$[H^+] = [OCl^-] = 5.95 \text{ x } 10^{-5} \text{ M}$

$[HOCl] = 0.10 - 5.9 \text{ x } 10^{-5} = 0.10 \text{ M}$

$$K_a = \frac{[H^+][OCl^-]}{[HOCl]} = \frac{(5.9 \text{ x } 10^{-5})(5.9 \text{ x } 10^{-5})}{0.10} = 3.5 \text{ x } 10^{-8}$$

Propanoic acid $HC_3H_5O_2 \rightleftharpoons H^+ + C_3H_5O_2^-$

Equilibrium concentrations:

$[H^+] = [C_3H_5O_2^-] = 1.4 \text{ x } 10^{-3} \text{ M}$

$[HC_3H_5O_2] = 0.15 - 1.4 \text{ x } 10^{-3} = 0.15 \text{ M}$

$$K_a = \frac{[H^+][C_3H_5O_2^-]}{[HC_3H_5O_2]} = \frac{(1.4 \text{ x } 10^{-3})(1.4 \text{ x } 10^{-3})}{0.15} = 1.3 \text{ x } 10^{-5}$$

Hydrocyanic acid $HCN \rightleftharpoons H^+ + CN^-$

Equilibrium concentrations:

$[H^+] = [CN^-] = 8.9 \text{ x } 10^{-6} \text{ M}$

$[HCN] = 0.20 - 8.9 \text{ x } 10^{-6} = 0.20 \text{ M}$

$$K_a = \frac{[H^+][CN^-]}{[HCN]} = \frac{(8.9 \text{ x } 10^{-6})^2}{0.20} = 4.0 \text{ x } 10^{-10}$$

45. **(a)** $HC_2H_3O_2 \rightleftharpoons H^+ + C_2H_3O_2^-$

Let x = molarity of $HC_2H_3O_2$ ionizing to establish equilibrium. Equilibrium concentrations are:

$[H^+] = [C_2H_3O_2^-] = x$

$[HC_2H_3O_2] = 0.25 - x = 0.25$

$K_a = \dfrac{[H^+]\,[C_2H_3O_2^-]}{[HC_2H_3O_2]} = \dfrac{x^2}{0.25} = 1.8 \times 10^{-5}$

$x^2 = (0.25)(1.8 \times 10^{-5})$

$x = \sqrt{(0.25)(1.8 \times 10^{-5})} = 2.1 \times 10^{-3} \text{ M} = [H^+]$

(b) $pH = -\log [H^+] = -\log (2.1 \times 10^{-3}) = 2.7$

(c) Percent ionization

$\dfrac{[H^+]}{[HC_2H_3O_2]}(100) = \left(\dfrac{2.1 \times 10^{-3}}{0.25}\right)(100) = 0.84\%$

46. $HA \rightleftharpoons H^+ + A^- \qquad 0.52\% = 0.0052$

$[H^+] = [A^-] = (1.0 \text{ M})(0.0052) = 5.2 \times 10^{-3} \text{ M}$

$[HA] = 1.0 \text{ M} - 0.0052 \text{ M} = 0.9948 \text{ M}$

$K_a = \dfrac{[H^+]\,[A^-]}{[HA]} = \dfrac{(5.2 \times 10^{-3})^2}{0.9948} = 2.7 \times 10^{-5}$

47. $HA \rightleftharpoons H^+ + A^- \qquad pH = 5 = -\log [H^+]$

$[H^+] = [A^-] = 1 \times 10^{-5} \text{ M}$

$[HA] = 0.15 \text{ M}$

$K_a = \dfrac{[H^+]\,[A^-]}{[HA]} = \dfrac{(1 \times 10^{-5})^2}{0.15} = 7 \times 10^{-10}$

48. $HC_2H_3O_2 \rightleftharpoons H^+ + C_2H_3O_2^-$

$K_a = \dfrac{[H^+][C_2H_3O_2^-]}{[HC_2H_3O_2]} = 1.8 \times 10^{-5}$

Let x = molarity of $HC_2H_3O_2$, which is ionized, to establish equilibrium. Equilibrium concentrations will be:

$[H^+] = [C_2H_3O_2^-] = x$

$[HC_2H_3O_2] = $ intitial concentration $- x$

Since K_a is small, the degree of ionization is small. Therefore, the approximation, initial concentration $- x = $ initial concentration, is valid.

(a) $[H^+] = [C_2H_3O_2^-] = x$ $[HC_2H_3O_2] = 1.0\ M$

$\dfrac{(x)(x)}{1.0} = 1.8 \times 10^{-5}$

$x^2 = (1.0)(1.8 \times 10^{-5})$

$x = \sqrt{1.8 \times 10^{-5}} = 4.2 \times 10^{-3}\ M$

$\left(\dfrac{4.2 \times 10^{-3}\ M}{1.0\ M}\right)(100) = 0.42\%$ ionized

$pH = -\log(4.2 \times 10^{-3}) = 2.4$

(b) $[HC_2H_3O_2] = 0.10\ M$

$\dfrac{(x)(x)}{0.10} = 1.8 \times 10^{-5}$

$x^2 = (0.10)(1.8 \times 10^{-5}) = 1.8 \times 10^{-6}$

$x = \sqrt{1.8 \times 10^{-6}} = 1.3 \times 10^{-3}\ M$

$\left(\dfrac{1.3 \times 10^{-3}\ M}{0.10\ M}\right)(100) = 1.3\%$ ionized

$pH = -\log(1.3 \times 10^{-3}) = 2.9$

(c) $[HC_2H_3O_2] = 0.010\ M$

$\dfrac{(x)(x)}{0.010} = 1.8 \times 10^{-5}$

$x^2 = (0.010)(1.8 \times 10^{-5}) = 1.8 \times 10^{-7}$

$x = \sqrt{1.8 \times 10^{-7}} = 4.2 \times 10^{-4}\ M$

$\left(\dfrac{4.2 \times 10^{-4}\ M}{0.010\ M}\right)(100) = 4.2\%$ ionized

$pH = -\log(4.2 \times 10^{-4}) = 3.4$

49. HA $\rightleftharpoons$ $H^+ + A^-$ $K_a = \dfrac{[H^+][A^-]}{[HA]}$

First, find the $[H^+]$. This calculated from the pH expression, $pH = -\log[H^+] = 3.7$. Enter -3.7 into the calculator and push the inverse (or 2nd) key, followed by the log key. This yields the $[H^+] = 2 \times 10^{-4}$.

$[H^+] = [A^-] = 2 \times 10^{-4}$ $[HA] = 0.37$

$K_a = \dfrac{[H^+][A^-]}{[HA]} = \dfrac{(2 \times 10^{-4})(2 \times 10^{-4})}{0.37} = 1 \times 10^{-7}$

50. See problem 49 for a discussion of calculating $[H^+]$ from pH.

$$HA \rightleftharpoons H^+ + A^- \qquad K_a = \frac{[H^+][A^-]}{[HA]}$$

pH = 2.89

$[H^+]$ = 1.3 x 10^{-3} = $[A^-]$ $\qquad$ [HA] = 0.23

$$K_a = \frac{[H^+][A^-]}{[HA]} = \frac{(1.3 \times 10^{-3})(1.3 \times 10^{-3})}{0.23} = 7.2 \times 10^{-6}$$

51. 6.0 M HCl yields $[H^+]$ = 6.0 M $\qquad$ (100% ionized)

pH = $-\log 6.0$ = -0.78

pOH = 14 $-$ pH = 14 $-$ (-0.78) = 14.78

$$[OH^-] = \frac{K_w}{[H^+]} = \frac{1 \times 10^{-14}}{6.0} = 1.7 \times 10^{-15}$$

52. pH + pOH = 14.0 $\qquad$ pOH = 14.0 $-$ pH

(a) $\quad$ 0.00010 M HCl $\quad$ $[H^+]$ = 0.00010 M = 1.0 x 10^{-4} M

pH = $-\log (1.0 \times 10^{-4})$ = 4.0

pOH = 14.0 $-$ 4.0 = 10.0

(b) $\quad$ 0.010 M NaOH $\quad$ $[OH^-]$ = 0.010 M = 1.0 x 10^{-2} M

pOH = $-\log (1.0 \times 10^{-2})$ = 2.0

pH = 14.0 $-$ 2.0 = 12.0

(c) $\quad$ 0.0025 M NaOH $\quad$ $[OH^-]$ = 2.5 x 10^{-3} M

pOH = $-\log (2.5 \times 10^{-3})$ = 2.6

pH = 14.0 $-$ 2.6 = 11.4

(d) $\quad$ HClO $\rightleftharpoons$ $\quad$ H^+ + ClO^-

$\quad$ 0.10 M $\qquad\quad$ x $\qquad\quad$ x

$$K_a = \frac{[H^+][ClO^-]}{[HClO]} = 3.5 \times 10^{-8}$$

$$\frac{(x)(x)}{0.10} = 3.5 \times 10^{-8}$$

x^2 = $(0.10)(3.5 \times 10^{-8})$

x = 5.9 x 10^{-5} = $[H^+]$

pH = $-\log (5.9 \times 10^{-5})$ = 4.2

pOH = 14.0 $-$ 4.2 = 9.8

(e) $\quad$ Fe(OH)$_2(s)$ $\rightleftharpoons$ Fe^{2+} + 2 OH^-

$\qquad\quad$ x $\qquad\qquad$ x $\qquad$ $2x$

$$K_{sp} = [Fe^{2+}][OH^-]^2 = (x)(2x)^2 = 8.0 \times 10^{-16}$$

$$4x^3 = 8.0 \times 10^{-16}$$

$$x = \sqrt[3]{\frac{8.0 \times 10^{-16}}{4}} = 5.8 \times 10^{-6}$$

$$[OH^-] = 2x = 2(5.8 \times 10^{-6}) = 1.2 \times 10^{-5}$$

$$pOH = -\log(1.2 \times 10^{-5}) = 4.9$$

$$pH = 14.0 - 4.9 = 9.1$$

53. Calculate the $[OH^-]$. $[OH^-] = \dfrac{K_w}{[H^+]}$

   (a) $[H^+] = 1.0 \times 10^{-4}$ $[OH^-] = \dfrac{1.0 \times 10^{-14}}{1.0 \times 10^{-4}} = 1.0 \times 10^{-10}$

   (b) $[H^+] = 2.8 \times 10^{-6}$ $[OH^-] = \dfrac{1.0 \times 10^{-14}}{2.8 \times 10^{-6}} = 3.6 \times 10^{-9}$

   (c) $[H^+] = 4.0 \times 10^{-9}$ $[OH^-] = \dfrac{1.0 \times 10^{-14}}{4.0 \times 10^{-9}} = 2.5 \times 10^{-6}$

54. Calculate $[H^+]$. $[H^+] = \dfrac{K_w}{[OH^-]}$

   (a) $[OH^-] = 6.0 \times 10^{-7}$ $[H^+] = \dfrac{1.0 \times 10^{-14}}{6.0 \times 10^{-7}} = 1.7 \times 10^{-8}$

   (b) $[OH^-] = 1 \times 10^{-8}$ $[H^+] = \dfrac{1 \times 10^{-14}}{1 \times 10^{-8}} = 1 \times 10^{-6}$

   (c) $[OH^-] = 4.5 \times 10^{-6}$ $[H^+] = \dfrac{1.0 \times 10^{-14}}{4.5 \times 10^{-6}} = 2.2 \times 10^{-9}$

55. The molar solubilities of the salts and their ions will be indicated below the formulas in the equilibrium equations.

   (a) $BaSO_4(s) \rightleftharpoons Ba^{2+} + SO_4^{2-}$

      3.9×10^{-5} 3.9×10^{-5} 3.9×10^{-5}

      $K_{sp} = [Ba^{2+}][SO_4^{2-}] = (3.9 \times 10^{-5})^2 = 1.5 \times 10^{-9}$

   (b) $Ag_2CrO_4(s) \rightleftharpoons 2\,Ag^+ + CrO_4^{2-}$

      7.8×10^{-5} $2(7.8 \times 10^{-5})$ 7.8×10^{-5}

      $K_{sp} = [Ag^+]^2[CrO_4^{2-}] = (2\{7.8 \times 10^{-5}\})^2(7.8 \times 10^{-5}) = 1.9 \times 10^{-12}$

   (c) $ZnS(s) \rightleftharpoons Zn^{2+} + S^{2-}$

      3.5×10^{-12} 3.5×10^{-12} 3.5×10^{-12}

      $K_{sp} = [Zn^{2+}][S^{2-}] = (3.5 \times 10^{-12})^2 = 1.2 \times 10^{-23}$

(d) $\quad Pb(IO_3)_2(s) \quad \rightleftharpoons \quad Pb^{2+} \quad + \quad 2\ IO_3^-$

$\quad\quad$ 4.0×10^{-5} $\quad\quad\quad\quad$ 4.0×10^{-5} $\quad\quad$ $2(4.0 \times 10^{-5})$

$\quad K_{sp} = [Pb^{2+}][IO_3^-]^2 = (4.0 \times 10^{-5})(2\{4.0 \times 10^{-5}\})^2 = 2.6 \times 10^{-13}$

(e) $\quad Bi_2S_3(s) \quad \rightleftharpoons \quad 2\ Bi^{3+} \quad + \quad 3\ S^{2-}$

$\quad\quad$ 4.9×10^{-15} $\quad\quad\quad$ $2(4.9 \times 10^{-15})$ $\quad$ $3(4.9 \times 10^{-15})$

$\quad K_{sp} = [Bi^{3+}]^2[S^{2-}]^3 = (2\{4.9 \times 10^{-15}\})^2(3\{4.9 \times 10^{-15}\})^3 = 3.1 \times 10^{-70}$

(f) $\quad \left(\dfrac{0.0019\ \text{g AgCl}}{L}\right)\left(\dfrac{1\ \text{mol}}{143.4\ \text{g}}\right) = 1.3 \times 10^{-5}\ M\ \ AgCl$

$\quad AgCl(s) \quad\quad \rightleftharpoons \quad\quad Ag^+ \quad + \quad Cl^-$

$\quad\quad$ 1.3×10^{-5} $\quad\quad\quad\quad$ 1.3×10^{-5} $\quad\quad$ 1.3×10^{-5}

$\quad K_{sp} = [Ag^+][Cl^-] = (1.3 \times 10^{-5})^2 = 1.7 \times 10^{-10}$

(g) $\quad \left(\dfrac{0.67\ \text{g CaSO}_4}{L}\right)\left(\dfrac{1\ \text{mol}}{136.1\ \text{g}}\right) = 4.9 \times 10^{-3}\ M\ \ CaSO_4$

$\quad CaSO_4(s) \quad\quad \rightleftharpoons \quad\quad Ca^{2+} \quad + \quad SO_4^{2-}$

$\quad\quad$ 4.9×10^{-3} $\quad\quad\quad\quad$ 4.9×10^{-3} $\quad\quad$ 4.9×10^{-3}

$\quad K_{sp} = [Ca^{2+}][SO_4^{2-}] = (4.9 \times 10^{-3})^2 = 2.4 \times 10^{-5}$

(h) $\quad \left(\dfrac{2.33 \times 10^{-4}\ \text{g Zn(OH)}_2}{L}\right)\left(\dfrac{1\ \text{mol}}{99.40\ \text{g}}\right) = 2.34 \times 10^{-6}\ M\ \ Zn(OH)_2$

$\quad Zn(OH)_2(s) \quad\quad \rightleftharpoons \quad\quad Zn^{2+} \quad + \quad 2\ OH^-$

$\quad\quad$ 2.34×10^{-6} $\quad\quad\quad\quad$ 2.34×10^{-6} $\quad\quad$ $2(2.34 \times 10^{-6})$

$\quad K_{sp} = [Zn^{2+}][OH^-]^2 = (2.34 \times 10^{-6})(2\{2.34 \times 10^{-6}\})^2 = 5.13 \times 10^{-17}$

(i) $\quad \left(\dfrac{6.73 \times 10^{-3}\ \text{g Ag}_3\text{PO}_4}{L}\right)\left(\dfrac{1\ \text{mol}}{418.7\ \text{g}}\right) = 1.61 \times 10^{-5}\ M\ \ Ag_3PO_4$

$\quad Ag_3PO_4(s) \quad\quad \rightleftharpoons \quad\quad 3\ Ag^+ \quad + \quad PO_4^{3-}$

$\quad\quad$ 1.61×10^{-5} $\quad\quad\quad\quad$ $3(1.61 \times 10^{-5})$ $\quad$ 1.61×10^{-5}

$\quad K_{sp} = [Ag^+]^3[PO_4^{3-}] = (3\{1.61 \times 10^{-5}\})^3(1.61 \times 10^{-5}) = 1.81 \times 10^{-18}$

56. The molar solubilities of the salts and their ions will be represented in terms of x below their formulas in the equilibrium equations.

(a) $\quad BaCO_3(s) \rightleftharpoons \quad\quad Ba^{2+} \quad + \quad CO_3^{2-}$

$\quad\quad\quad$ x $\quad\quad\quad\quad\quad\quad$ x $\quad\quad\quad$ x

$\quad K_{sp} = [Ba^{2+}][CO_3^{2-}] = x^2 = 2.0 \times 10^{-9}$

$\quad x = \sqrt{2.0 \times 10^{-9}} = 4.5 \times 10^{-5}\ M$

(b) $AlPO_4(s) \rightleftharpoons$ Al^{3+} + PO_4^{3-}

$\quad x \qquad\qquad\qquad x \qquad\quad x$

$K_{sp} = [Al^{3+}][PO_4^{3-}] = x^2 = 5.8 \times 10^{-19}$

$x = \sqrt{5.8 \times 10^{-19}} = 7.6 \times 10^{-10}$ M

(c) $Ag_2SO_4(s) \rightleftharpoons$ $2\,Ag^+$ + SO_4^{2-}

$\quad x \qquad\qquad\qquad 2x \qquad\quad x$

$K_{sp} = [Ag^+]^2[SO_4^{2-}] = (2x)^2(x) = 4x^3 = 1.5 \times 10^{-5}$

$x = \sqrt[3]{\dfrac{1.5 \times 10^{-5}}{4}} = 1.6 \times 10^{-2}$ M

(d) $Mg(OH)_2(s) \rightleftharpoons$ Mg^{2+} + $2\,OH^-$

$\quad x \qquad\qquad\qquad x \qquad\quad 2x$

$K_{sp} = [Mg^{2+}][OH^-]^2 = (x)(2x)^2 = 4x^3 = 7.1 \times 10^{-12}$

$x = \sqrt[3]{\dfrac{7.1 \times 10^{-12}}{4}} = 1.2 \times 10^{-4}$ M

57. (a) $\left(\dfrac{4.5 \times 10^{-5} \text{ mol BaCO}_3}{L}\right)(0.100\text{ L})\left(\dfrac{197.3\text{ g}}{\text{mol}}\right) = 8.9 \times 10^{-4}$ g $BaCO_3$

(b) $\left(\dfrac{7.6 \times 10^{-10} \text{ mol AlPO}_4}{L}\right)(0.100\text{ L})\left(\dfrac{122.0\text{ g}}{\text{mol}}\right) = 9.3 \times 10^{-9}$ g $AlPO_4$

(c) $\left(\dfrac{1.6 \times 10^{-2}\text{mol Ag}_2\text{SO}_4}{L}\right)(0.100\text{ L})\left(\dfrac{311.9\text{ g}}{\text{mol}}\right) = 0.50$ g Ag_2SO_4

(d) $\left(\dfrac{1.2 \times 10^{-4} \text{ mol Mg(OH)}_2}{L}\right)(0.100\text{ L})\left(\dfrac{58.33\text{ g}}{\text{mol}}\right) = 7.0 \times 10^{-4}$ g $Mg(OH)_2$

58. Let x = M CaF_2 dissolving

$CaF_2(s) \rightleftharpoons$ Ca^{2+} + $2\,F^-$

$\quad x \qquad\qquad\qquad x \qquad\quad 2x$

(a) $K_{sp} = [Ca^{2+}][F^-]^2 = (x)(2x)^2 = 4x^3 = 3.9 \times 10^{-11}$

$x = \sqrt[3]{\dfrac{3.9 \times 10^{-11}}{4}} = 2.1 \times 10^{-4}$ M CaF_2 dissolving

$\left(\dfrac{2.1 \times 10^{-4} \text{ mol CaF}_2}{L}\right)\left(\dfrac{1\text{ mol Ca}^{2+}}{1\text{ mol CaF}_2}\right) = 2.1 \times 10^{-4}$ M Ca^{2+}

$\left(\dfrac{2.1 \times 10^{-4} \text{ mol CaF}_2}{L}\right)\left(\dfrac{2\text{ mol F}^-}{1\text{ mol CaF}_2}\right) = 4.2 \times 10^{-4}$ M F^-

(b) $\left(\dfrac{2.1 \times 10^{-4} \text{ mol CaF}_2}{L}\right)(0.500 \text{ L})\left(\dfrac{78.08 \text{ g}}{\text{mol}}\right) = 8.2 \times 10^{-3} \text{ g CaF}_2$

59. The molar concentrations of ions, after mixing, are calculated and these concentrations are substituted into the equilibrium expression. The value obtained is compared to the K_{sp} of the salt. If the value is greater than the K_{sp}, precipitation occurs. If the value is less than the K_{sp}, no precipitation occurs.

(a) 100. mL 0.010 M Na_2SO_4 $\longrightarrow$ 100. mL 0.010 M SO_4^{2-}

100. mL 0.001 M $Pb(NO_3)_2$ $\longrightarrow$ 100. mL 0.001 M Pb^{2+}

Volume after mixing = 200. mL

Concentrations after mixing:

SO_4^{2-} = 0.0050 M Pb^{2+} = 0.0005 M

$[Pb^{2+}][SO_4^{2-}] = (5 \times 10^{-3})(5 \times 10^{-4}) = 2.5 \times 10^{-6}$

$K_{sp} = 1.3 \times 10^{-8}$ which is less than 2.5×10^{-6}, therefore, precipitation occurs.

(b) 50.0 mL 1.0×10^{-4} M $AgNO_3$ $\longrightarrow$ 50.0 mL 1.0×10^{-4} M Ag^+

100. mL 1.0×10^{-4} M NaCl $\longrightarrow$ 100. mL 1.0×10^{-4} M Cl^-

Volume after mixing = 150. mL

Concentrations after mixing:

$(1.0 \times 10^{-4} \text{ M } Ag^+)\left(\dfrac{50.0 \text{ mL}}{150. \text{ mL}}\right) = 3.3 \times 10^{-5}$ M Ag^+

$(1.0 \times 10^{-4} \text{ M } Cl^-)\left(\dfrac{100. \text{ mL}}{150. \text{ mL}}\right) = 6.7 \times 10^{-5}$ M Cl^-

$[Ag^+][Cl^-] = (3.3 \times 10^{-5})(6.7 \times 10^{-5}) = 2.2 \times 10^{-9}$

$K_{sp} = 1.7 \times 10^{-10}$ which is less than 2.2×10^{-9}, therefore precipitation occurs.

(c) $\left(\dfrac{1.0 \text{ g Ca(NO}_3)_2}{0.150 \text{ L}}\right)\left(\dfrac{1 \text{ mol}}{164.1 \text{ g}}\right)\left(\dfrac{1 \text{ mol Ca}^{2+}}{1 \text{ mol Ca(NO}_3)_2}\right) = 0.041$ M Ca^{2+}

250 mL 0.01 M NaOH $\longrightarrow$ 250 mL 0.01 M OH^-

Final volume = 4.0×10^2 mL

Concentrations after mixing:

$(0.041 \text{ M } Ca^{2+})\left(\dfrac{150 \text{ mL}}{4.0 \times 10^2 \text{ mL}}\right) = 0.015$ M Ca^{2+}

$(0.01 \text{ M } OH^-)\left(\dfrac{250 \text{ mL}}{4.0 \times 10^2 \text{ mL}}\right) = 0.0063$ M OH^-

$[Ca^{2+}][OH^-]^2 = (0.015)(0.0063)^2 = 6.0 \times 10^{-7}$

$K_{sp} = 1.3 \times 10^{-6}$ which is greater than 6.0×10^{-7}, therefore, no precipitation occurs.

60. With a known Ba^{2+} concentration, the SO_4^{2-} concentration can be calculated using the K_{sp} value.

$$K_{sp} = [Ba^{2+}][SO_4^{2-}] = 1.5 \times 10^{-9} \quad Ba^{2+} = 0.50 \text{ M}$$

(a) $\quad [SO_4^{2-}] = \dfrac{K_{sp}}{[Ba^{2+}]} = \dfrac{1.5 \times 10^{-9}}{0.050} = 3.0 \times 10^{-8} \text{ M } SO_4^{2-}$ in solution

(b) $\quad$ M SO_4^{2-} = M $BaSO_4$ dissolving

$$\left(\frac{3.0 \times 10^{-8} \text{ mol } BaSO_4}{L}\right)(0.100 \text{ L})\left(\frac{233.4 \text{ g}}{\text{mol}}\right) = 7.0 \times 10^{-7} \text{ g } BaSO_4 \text{ remain in solution}$$

61. $\quad [Ba^{2+}][SO_4^{2-}] = 1.5 \times 10^{-9} \quad [Sr^{2+}][SO_4^{2-}] = 3.5 \times 10^{-7}$

Both cations are present in equal concentrations (0.10 M). Therefore, as SO_4^{2-} is added, the K_{sp} of $BaSO_4$ will be exceeded before that of $SrSO_4$. $BaSO_4$ precipitates first.

62. The concentration of $Br^- = 0.10$ M in 1.0 L of 0.10 M NaBr.

$$K_{sp} = [Ag^+][Br^-] = 5.0 \times 10^{-13}$$

(a) $\quad [Ag^+] = \dfrac{5.0 \times 10^{-13}}{[Br^-]} = \dfrac{5.0 \times 10^{-13}}{0.10} = 5.0 \times 10^{-12} \text{ M}$

$$\left(\frac{5.0 \times 10^{-12} \text{ mol } Ag^+}{L}\right)\left(\frac{1 \text{ mol } AgBr}{1 \text{ mol } Ag^+}\right)(1.0 \text{ L}) = 5.0 \times 10^{-12} \text{ mol } AgBr \text{ in solution}$$

(b) $\quad \left(\dfrac{0.10 \text{ mol } MgBr_2}{L}\right)\left(\dfrac{2 \text{ mol } Br^-}{1 \text{ mol } MgBr_2}\right) = \left(\dfrac{0.20 \text{ mol } Br^-}{L}\right) = 0.20 \text{ M } Br^-$

$$[Ag^+] = \dfrac{5.0 \times 10^{-13}}{[Br^-]} = \dfrac{5.0 \times 10^{-13}}{0.20} = 2.5 \times 10^{-12} \text{ M}$$

$$\left(\frac{2.5 \times 10^{-12} \text{ mol } Ag^+}{L}\right)\left(\frac{1 \text{ mol } AgBr \text{ in solution}}{1 \text{ mol } Ag^+}\right)(1.0 \text{ L}) = 2.5 \times 10^{-12} \text{ mol } AgBr \text{ in solution}$$

63. If $[Pb^{2+}][Cl^-]^2$ exceeds the K_{sp}, precipitation will occur.

$$K_{sp} = [Pb^{2+}][Cl^-]^2 = 2.0 \times 10^{-5}$$

0.050 M $Pb(NO_3)_2 \longrightarrow$ 0.050 M Pb^{2+}

0.010 M NaCl $\longrightarrow$ 0.010 M Cl^-

$$(0.050)(0.010)^2 = 5.0 \times 10^{-6}$$

$[Pb^{2+}][Cl^-]^2$ is smaller than the K_{sp} value. Therefore, no precipitate of $PbCl_2$ will form.

64. $2 SO_2(g) + O_2(g) \rightleftharpoons 2 SO_3(g)$

$$K_{eq} = \frac{[SO_3]^2}{[SO_2]^2[O_2]} = \frac{(11.0)^2}{(4.20)^2(0.6 \times 10^{-3})} = 1 \times 10^4$$

65. $(0.048 \text{ g BaF}_2)\left(\frac{1 \text{ mol}}{175.3 \text{ g}}\right) = 2.7 \times 10^{-4} \text{ mol BaF}_2$

$\left(\frac{2.7 \times 10^{-4} \text{ mol}}{0.015 \text{ L}}\right) = 1.8 \times 10^{-2} \text{ M BaF}_2 \text{ dissolving}$

$\text{BaF}_2(s) \rightleftharpoons \text{Ba}^{2+} + 2 \text{ F}^-$

$1.8 \times 10^{-2} \quad 1.8 \times 10^{-2} \quad 2(1.8 \times 10^{-2})$

$K_{sp} = [\text{Ba}^{2+}][\text{F}^-]^2 = (1.8 \times 10^{-2})(3.6 \times 10^{-2})^2 = 2.3 \times 10^{-5}$

66. $\text{N}_2 + 3 \text{ H}_2 \rightleftharpoons 2 \text{ NH}_3$

$K_{eq} = \frac{[\text{NH}_3]^2}{[\text{N}_2][\text{H}_2]^3} = 4.0$

$4.0 = \frac{x^2}{(2.0)(2.0)^3} \qquad x^2 = 64$

$x = 8.0 \text{ M} = [\text{NH}_3]$

67. Total volume of mixture $= 40.0 \text{ mL}$

$$[\text{Sr}^{2+}] = \frac{(1.0 \times 10^{-3} \text{ M })(0.025 \text{ L})}{0.040 \text{ L}} = 6.3 \times 10^{-4} \text{ M}$$

$$[\text{SO}_4{}^{2-}] = \frac{(2.0 \times 10^{-3} \text{ M})(0.015 \text{ L})}{0.040 \text{ L}} = 7.5 \times 10^{-4} \text{ M}$$

$[\text{Sr}^{2+}][\text{SO}_4{}^{2-}] = (6.3 \times 10^{-4})(7.5 \times 10^{-4}) = 4.7 \times 10^{-7}$

$4.7 \times 10^{-7} < 7.6 \times 10^{-7}$ no precipitation should occur.

68. $\left(\frac{3.04 \times 10^{-7} \text{ g Hg}_2\text{I}_2}{\text{L}}\right)\left(\frac{1 \text{ mol}}{655.0 \text{ g}}\right) = 4.64 \times 10^{-10} \text{ M Hg}_2\text{I}_2 \text{ dissolving}$

$K_{sp} = [\text{Hg}_2{}^{2+}][\text{I}^-]^2 = (4.64 \times 10^{-10})(2\{4.64 \times 10^{-10}\})^2 = 4.00 \times 10^{-28}$

69. $\text{HC}_2\text{H}_3\text{O}_2 \rightleftharpoons \text{H}^+ + \text{C}_2\text{H}_3\text{O}_2{}^-$

$K_a = \frac{[\text{H}^+][\text{C}_2\text{H}_3\text{O}_2{}^-]}{[\text{HC}_2\text{H}_3\text{O}_2]} = 1.8 \times 10^{-5}$

$[\text{H}^+] = K_a\left(\frac{[\text{HC}_2\text{H}_3\text{O}_2]}{[\text{C}_2\text{H}_3\text{O}_2{}^-]}\right) \qquad [\text{HC}_2\text{H}_3\text{O}_2] = 0.20 \text{ M}$

(a) $[\text{C}_2\text{H}_3\text{O}_2{}^-] = 0.10 \text{ M} \qquad [\text{H}^+] = (1.8 \times 10^{-5})\left(\frac{0.20}{0.10}\right) = 3.6 \times 10^{-5} \text{ M}$

$\text{pH} = -\log(3.6 \times 10^{-5}) = 4.4$

(b) $[\text{C}_2\text{H}_3\text{O}_2{}^-] = 0.20 \text{ M} \qquad [\text{H}^+] = (1.8 \times 10^{-5})\left(\frac{0.20}{0.20}\right) = 1.8 \times 10^{-5} \text{ M}$

$\text{pH} = -\log(1.8 \times 10^{-5}) = 4.7$

70. **(a)** Initially, the solution of NaCl is neutral. $[H^+] = 1.0 \times 10^{-7}$

$pH = -\log(1.0 \times 10^{-7}) = 7.0$

Dilution: $(1.0 \text{ M})(1.0 \text{ mL}) = M_2(51 \text{ mL})$

$M_2 = \dfrac{(1.0 \text{ M})(1.0 \text{ mL})}{51 \text{ mL}} = 0.020 \text{ M}$

The diluted $[H^+] = 2.0 \times 10^{-2}$

$pH = -\log(2.0 \times 10^{-2}) = 1.7$

Change in pH $= 7.0 - 1.7 = 5.3$ units

(b) Initially, $[H^+] = 1.8 \times 10^{-5}$

$pH = -\log(1.8 \times 10^{-5}) = 4.74$

Final $[H^+] = 1.9 \times 10^{-5}$

$pH = -\log(1.9 \times 10^{-5}) = 4.72$

Change in pH $= 4.74 - 4.72 = 0.02$ units in the buffered solution

OXIDATION-REDUCTION

1. (a) Iodine is oxidized. Its oxidation number increases from 0 to +5.

    (b) Chlorine is reduced. Its oxidation number decreases from 0 to −1.

2. The metal which is higher on the list is more reactive.

    (a) Al   (b) Ba   (c) Ni

3. If the free element is higher on the list than the ion with which it is paired, the reaction occurs.

    (a) Yes. $Zn + Cu^{2+} \longrightarrow Zn^{2+} + Cu$

    (b) No

    (c) Yes. $Sn + 2\,Ag^+ \longrightarrow Sn^{2+} + 2\,Ag$

    (d) No

    (e) Yes. $Ba + FeCl_2 \longrightarrow BaCl_2 + Fe$

    (f) No

    (g) Yes. $Ni + Hg(NO_3)_2 \longrightarrow Ni(NO_3)_2 + Hg$

    (h) Yes. $2\,Al + 3\,CuSO_4 \longrightarrow Al_2(SO_4)_3 + 3\,Cu$

4. (a) $2\,Al + Fe_2O_3 \longrightarrow Al_2O_3 + 2\,Fe$

    (b) Al is above Fe in the activity series, which indicates Al is more active than Fe.

    (c) No. Iron is less active than aluminum and will not displace the aluminum ion from its compounds.

    (d) Yes. Aluminum is above chromium in the activity series and will displace Cr^{3+} from its compounds.

5. (a) $2\,Al + 6\,HCl \longrightarrow 2\,AlCl_3 + 3\,H_2$

    $2\,Al + 3\,H_2SO_4 \longrightarrow Al_2(SO_4)_3 + 3\,H_2$

    (b) $2\,Cr + 6\,HCl \longrightarrow 2\,CrCl_3 + 3\,H_2$

    $2\,Cr + 3\,H_2SO_4 \longrightarrow Cr_2(SO_4)_3 + 3\,H_2$

    (c) $Au + HCl \longrightarrow$ no reaction

    $Au + H_2SO_4 \longrightarrow$ no reaction

    (d) $Fe + 2\,HCl \longrightarrow FeCl_2 + H_2$

    $Fe + H_2SO_4 \longrightarrow FeSO_4 + H_2$

(e) $Cu + HCl \longrightarrow$ no reaction

     $Cu + H_2SO_4 \longrightarrow$ no reaction

(f) $Mg + 2\,HCl \longrightarrow MgCl_2 + H_2$

     $Mg + H_2SO_4 \longrightarrow MgSO_4 + H_2$

(g) $Hg + HCl \longrightarrow$ no reaction

     $Hg + H_2SO_4 \longrightarrow$ no reaction

(h) $Zn + 2\,HCl \longrightarrow ZnCl_2 + H_2$

     $Zn + H_2SO_4 \longrightarrow ZnSO_4 + H_2$

6. (a)   Oxidation occurs at the anode.   The reaction is

$$2\,Cl^- \longrightarrow Cl_2 + 2\,e^-$$

(b)   Reduction occurs at the cathode.   The reaction is

$$Ni^{2+} + 2\,e^- \longrightarrow Ni$$

(c)   The net chemical reaction is

$$Ni^{2+} + 2\,Cl^- \xrightarrow[\text{energy}]{\text{electrical}} Ni + Cl_2$$

7. In Figure 18.2, electrical energy is causing chemical reactions to occur.   In Figure 18.3, chemical reactions are used to produce electrical energy.

8. (a)   It would not be possible to monitor the voltage produced, but the reactions in the cell would still occur.

(b)   If the salt bridge were removed, the reaction would stop.   Ions must be mobile to maintain an electrical neutrality of ions in solution.   The two solutions would be isolated with no complete electrical circuit.

9. The oxidation number of the underlined element is indicated by the number following the formula.

(a)	$\underline{N}aCl$ +1	(e)	$H_2\underline{S}O_3$ +4	(i)	$\underline{N}H_3$ -3
(b)	$Fe\underline{Cl}_3$ -1	(f)	$\underline{N}H_4Cl$ -3	(j)	$K\underline{Cl}O_3$ +5
(c)	$\underline{Pb}O_2$ +4	(g)	$K\underline{Mn}O_4$ +7	(k)	$K_2\underline{Cr}O_4$ +6
(d)	$Na\underline{N}O_3$ +5	(h)	$\underline{I}_2$ 0	(l)	$K_2\underline{Cr}_2O_7$ +6

10. The oxidation number of the underlined element in each formula, is indicated by the number following the formula.

(a)	$\underline{S}^{2-}$ -2	(d)	$\underline{Mn}O_4^-$ +7	(g)	$\underline{As}O_4^{3-}$ +5
(b)	$\underline{N}O_2^-$ +3	(e)	$\underline{Bi}^{3+}$ +3	(h)	$Fe(\underline{O}H)_3$ -2
(c)	$Na_2\underline{O}_2$ -1	(f)	$\underline{O}_2$ 0	(i)	$\underline{I}O_3^-$ +5

11. Half–reactions

Balanced half–reaction	Changing Element	Type of reaction
(a) $Zn^{2+} + 2\ e^- \longrightarrow Zn$	Zn	reduction
(b) $2\ Br^- \longrightarrow Br_2 + 2\ e^-$	Br	oxidation
(c) $MnO_4^- + 8\ H^+ + 5\ e^- \longrightarrow Mn^{2+} + 4\ H_2O$	Mn	reduction
(d) $Ni \longrightarrow Ni^{2+} + 2\ e^-$	Ni	oxidation
(e) $SO_3^{2-} + H_2O \longrightarrow SO_4^{2-} + 2\ H^+ + 2\ e^-$	S	oxidation
(f) $NO_3^- + 4\ H^+ + 3\ e^- \longrightarrow NO + 2\ H_2O$	N	reduction
(g) $S_2O_4^{2-} + 2\ H_2O \longrightarrow 2\ SO_3^{2-} + 4\ H^+ + 2\ e^-$	S	oxidation
(h) $Fe^{2+} \longrightarrow Fe^{3+} + 1\ e^-$	Fe	oxidation

12. (1) $Cr + HCl \longrightarrow CrCl_3 + H_2$

      (a) Cr is oxidized, H is reduced

      (b) HCl is the oxidizing agent, Cr the reducing agent

(2) $SO_4^{2-} + I^- + H^+ \longrightarrow H_2S + I_2 + H_2O$

      (a) I^- is oxidized, S is reduced

      (b) SO_4^{2-} is the oxidizing agent, I^- the reducing agent

(3) $AsH_3 + Ag^+ + H_2O \longrightarrow H_3AsO_4 + Ag + H^+$

      (a) As is oxidized, Ag^+ is reduced

      (b) Ag^+ is the oxidizing agent, AsH_3 the reducing agent

(4) $Cl_2 + NaBr \longrightarrow NaCl + Br_2$

      (a) Br is oxidized, Cl is reduced

      (b) Cl_2 is the oxidizing agent, NaBr the reducing agent

13. Balancing oxidation–reduction equations

    (a) $Zn + S \longrightarrow ZnS$

ox $Zn^0 \longrightarrow Zn^{2+} + 2\ e^-$ Add half reactions

red $\underline{S^0 + 2\ e^- \longrightarrow S^{2-}}$ the $2e^-$ cancel

                 $Zn + S \longrightarrow ZnS$

(b)
$$AgNO_3 + Pb \longrightarrow Pb(NO_3)_2 + Ag$$

ox $\qquad Pb^0 \longrightarrow Pb^{2+} + 2\ e^-$

red $\qquad \underline{Ag^+ + 1\ e^- \longrightarrow Ag^0}$ $\qquad$ Multiply by 2, add the half

$\qquad Pb + 2\ Ag^+ \longrightarrow Pb^{2+} + 2\ Ag$ $\quad$ reactions, the 2 e⁻ cancel

Transfer the coefficients to the original equation and complete the balancing by inspection.

$$2\ AgNO_3 + Pb \longrightarrow Pb(NO_3)_2 + 2\ Ag$$

(c)
$$Fe_2O_3 + CO \longrightarrow Fe + CO_2$$

ox $\qquad C^{2+} \longrightarrow C^{4+} + 2\ e^-$ $\qquad$ Multiply by 3

red $\qquad \underline{Fe^{3+} + 3\ e^- \longrightarrow Fe^0}$ $\qquad$ Multiply by 2, add, the 6 e⁻ cancel

$\qquad 3\ C^{2+} + 2\ Fe^{3+} \longrightarrow 3\ C^{4+} + Fe$

Transfer the coefficients to the original equation (the coefficient 2 in front of the Fe^{3+} becomes the subscript 2 in Fe_2O_3). Complete the balancing by inspection.

$$Fe_2O_3 + 3\ CO \longrightarrow 2\ Fe + 3\ CO_2$$

(d) $\qquad H_2S + HNO_3 \longrightarrow S + NO + H_2O$

$S^{2-} \longrightarrow S^0 + 2\ e^-$ $\qquad$ Multiply by 3

$\underline{N^{5+} + 3\ e^- \longrightarrow N^{2+}}$ $\qquad$ Multiply by 2, add, the 6 e⁻ cancel

$3\ S^{2-} + 2\ N^{5+} \longrightarrow 3\ S + 2\ N^{2+}$

Transfer the coefficients to the original equations and complete the balancing by inspection.

$$3\ H_2S + 2\ HNO_3 \longrightarrow 3\ S + 2\ NO + 4\ H_2O$$

(e) $\qquad MnO_2 + HBr \longrightarrow MnBr_2 + Br_2 + H_2O$

$Br^- \longrightarrow Br^0 + 1\ e^-$ $\qquad$ Multiply by 2

$\underline{Mn^{4+} + 2\ e^- \longrightarrow Mn^{2+}}$ $\qquad$ Add equations and the 2 e⁻ cancel

$Mn^{4+} + 2\ Br^- \longrightarrow Mn^{2+} + 2\ Br$

Transfer the coefficients to the original equation. The coefficient 2 in front of the Br^- becomes the subscript 2 in Br_2. Also, 2 more Br^- ions are required to account for the 2 Br^- ions that do not change oxidation numbers. These 2 are part of the compound $MnBr_2$.

$$MnO_2 + 4\ HBr \longrightarrow MnBr_2 + Br_2 + 2\ H_2O$$

(f) $Cl_2 + KOH \longrightarrow KCl + KClO_3$

$Cl^0 \longrightarrow Cl^{5+} + 5\ e^-$

$\underline{Cl^0 + e^- \longrightarrow Cl^-}$ Multiply by 5, add, the 5 e^- cancel

$3\ Cl_2 \longrightarrow Cl^{5+} + 5\ Cl^-$

Tranfer the coefficients to the original equations and complete the balancing by inspection.

$3\ Cl_2 + 6\ KOH \longrightarrow KClO_3 + 5\ KCl + 3\ H_2O$

(g) $Ag + HNO_3 \longrightarrow AgNO_3 + NO + H_2O$

$Ag^0 \longrightarrow Ag^+ + e^-$ Multiply by 3, add, the

                                 3 e^- cancel

$\underline{N^{5+} + 3\ e^- \longrightarrow N^{2+}}$

$3\ Ag + N^{5+} \longrightarrow 3\ Ag^+ + N^{2+}$

Transfer the coefficients to the original equation and complete the balancing by inspection.

$3\ Ag + 4\ HNO_3 \longrightarrow 3\ AgNO_3 + NO + 2\ H_2O$

(h) $CuO + NH_3 \longrightarrow N_2 + Cu + H_2O$

$N^{3-} \longrightarrow N^0 + 3\ e^-$ Multiply by 2

$\underline{Cu^{2+} + 2\ e^- \longrightarrow Cu^0}$ Multiply by 3, add, the 6 e^- cancel

$2\ N^{3-} + 3\ Cu^{2+} \longrightarrow N_2 + 3\ Cu$

Transfer the coefficients to the original equation and complete the balancing by inspection.

$3\ CuO + 2\ NH_3 \longrightarrow N_2 + 3\ Cu + H_2O$

(i) $PbO_2 + Sb + NaOH \longrightarrow PbO + NaSbO_2 + H_2O$

$Sb^0 \longrightarrow Sb^{3+} + 3\ e^-$ Mutiply by 2

$\underline{Pb^{4+} + 2\ e^- \longrightarrow 2\ Pb^{2+}}$ Multiply by 3, add, the 6 e^- cancel

$2\ Sb + 3\ Pb^{4+} \longrightarrow 2\ Sb^{3+} + 3\ Pb^{2+}$

Transfer the coefficients to the original equation and complete the balancing by inspection.

$3\ PbO_2 + 2\ Sb + 2\ NaOH \longrightarrow 3\ PbO + 2\ NaSbO_2 + H_2O$

(j) $\quad H_2O_2 \; + \; KMnO_4 \; + \; H_2SO_4 \; \longrightarrow \; O_2 \; + \; MnSO_4 + K_2SO_4 + H_2O$

$O_2{}^{2-} \; \longrightarrow \; O_2{}^0 \; + \; 2\;e^- \qquad$ Multiply by 5

$\underline{Mn^{7+} \; + \; 5\;e^- \; \longrightarrow \; Mn^{2+}} \qquad$ Multiply by 2, add, the 10 e^- cancel

$5\;O_2{}^{2-} \; + \; 2\;Mn^{7+} \; \longrightarrow \; 5\;O_2 \; + \; 2\;Mn^{2+}$

Transfer the coefficients to the original equation and complete the balancing by inspection.

$5\;H_2O_2 \; + \; 2\;KMnO_4 \; + \; 3\;H_2SO_4 \; \longrightarrow \; 5\;O_2 + 2\;MnSO_4 + K_2SO_4 + 8\;H_2O$

14. (a) $\quad Zn \; + \; NO_3{}^- \; \longrightarrow \; Zn^{2+} \; + \; NH_4{}^+$

Step 1   Write half–reaction equations. Balance except H and O.

$Zn \; \longrightarrow \; Zn^{2+}$

$NO_3{}^- \; \longrightarrow \; NH_4{}^+$

Step 2   Balance H and O using H_2O and H^+

$Zn \; \longrightarrow \; Zn^{2+}$

$10\;H^+ \; + \; NO_3{}^- \; \longrightarrow \; NH_4{}^+ \; + \; 3\;H_2O$

Step 3   Balance electrically with electrons

$Zn \; \longrightarrow \; Zn^{2+} \; + \; 2\;e^-$

$10\;H^+ \; + \; NO_3{}^- \; + \; 8\;e^- \; \longrightarrow \; NH_4{}^+ \; + \; 3\;H_2O$

Step 4   Equalize the loss and gain of electrons

$4\;(Zn \; \longrightarrow \; Zn^{2+} \; + \; 2\;e^-)$

$10\;H^+ \; + \; NO_3{}^- \; + \; 8\;e^- \; \longrightarrow \; NH_4{}^+ \; + \; H_2O$

Step 5   Add the half reactions

$10\;H^+ + 4\;Zn + NO_3{}^- \; \longrightarrow \; 4\;Zn^{2+} + NH_4{}^+ + 3\;H_2O$

(b) $\quad NO_3{}^- \; + \; S \; \longrightarrow \; NO_2 \; + \; SO_4{}^{2-}$

Step 1   Write half–reactions equations. Balance except H and O.

$S \; \longrightarrow \; SO_4{}^{2-}$

$NO_3{}^- \; \longrightarrow \; NO_2$

Step 2   Balance H and O using H_2O and H^+

$4\;H_2O \; + \; S \; \longrightarrow \; SO_4{}^{2-} \; + \; 8\;H^+$

$2\;H^+ \; + \; NO_3{}^- \; \longrightarrow \; NO_2 \; + \; H_2O$

Step 3   Balance electrically with electrons

$$4\ H_2O\ +\ S\ \longrightarrow\ SO_4^{2-}\ +\ 8\ H^+\ +\ 6\ e^-$$

$$2\ H^+\ +\ NO_3^-\ +\ e^-\ \longrightarrow\ NO_2\ +\ H_2O$$

Steps 4 and 5   Equalize loss and gain of electrons; add half–reactions

$$4\ H_2O\ +\ S\ \longrightarrow\ SO_4^{2-}\ +\ 8\ H^+\ +\ 6\ e^-$$

$$\underline{6\ (2\ H^+\ +\ NO_3^-\ +\ e^-\ \longrightarrow\ NO_2\ +\ H_2O)}$$

$$4\ H^+\ +\ S\ +\ 6\ NO_3^-\ \longrightarrow\ 6\ NO_2\ +\ SO_4^{2-}\ +\ 2\ H_2O$$

(c) $PH_3\ +\ I_2\ \longrightarrow\ H_3PO_2\ +\ I^-$

Step 1   Write half–reaction equations. Balance except H and O.

$$PH_3\ \longrightarrow\ H_3PO_2$$

$$I_2\ \longrightarrow\ 2\ I^-$$

Step 2   Balance H and O using H_2O and H^+

$$2\ H_2O\ +\ PH_3\ \longrightarrow\ H_3PO_2\ +\ 4\ H^+$$

$$I_2\ \longrightarrow\ 2\ I^-$$

Step 3   Balance electrically with electrons

$$2\ H_2O\ +\ PH_3\ \longrightarrow\ H_3PO_2\ +\ 4\ H^+\ +\ 4\ e^-$$

$$I_2\ +\ 2\ e^-\ \longrightarrow\ 2\ I^-$$

Steps 4 and 5   Equalize the loss and gain of electrons; add half–reactions

$$2\ H_2O\ +\ PH_3\ \longrightarrow\ H_3PO_2\ +\ 4\ H^+\ +\ 4\ e^-$$

$$\underline{2\ (I_2\ +\ 2\ e^-\ \longrightarrow\ 2\ I^-)}$$

$$PH_3\ +\ 2\ H_2O\ +\ 2\ I_2\ \longrightarrow\ H_3PO_2\ +\ 4\ I^-\ +\ 4\ H^+$$

(d) $Cu\ +\ NO_3^-\ \longrightarrow\ Cu^{2+}\ +\ NO$

Step 1   Write half–reaction equations. Balance except H and O.

$$Cu\ \longrightarrow\ Cu^{2+}$$

$$NO_3^-\ \longrightarrow\ NO$$

Step 2   Balance H and O using H_2O and H^+

$$Cu\ \longrightarrow\ Cu^{2+}$$

$$4\ H^+\ +\ NO_3^-\ \longrightarrow\ NO\ +\ 2\ H_2O$$

Step 3   Balance electrically with electrons

$$Cu\ \longrightarrow\ Cu^{2+}\ +\ 2\ e^-$$

$$4\ H^+\ +\ NO_3^-\ +\ 3\ e^-\ \longrightarrow\ NO\ +\ 2\ H_2O$$

Steps 4 and 5   Equalize the loss and gain of electrons; add half-reactions

$$3 \; (Cu \longrightarrow Cu^{2+} + 2 \; e^-)$$

$$\underline{2 \; (4 \; H^+ + NO_3^- + 3 \; e^- \longrightarrow NO + 2 \; H_2O)}$$

$$3 \; Cu + 8 \; H^+ + 2 \; NO_3^- \longrightarrow 3 \; Cu^{2+} + 2 \; NO + 4 \; H_2O$$

(e) $ClO_3^- + I^- \longrightarrow I_2 + Cl^-$

Step 1   Write half–reaction equations. Balance except H and O.

$$2 \; I^- \longrightarrow I_2$$

$$ClO_3^- \longrightarrow Cl^-$$

Step 2   Balance H and O using H_2O and H^+

$$2 \; I^- \longrightarrow I_2$$

$$6 \; H^+ + ClO_3^- \longrightarrow Cl^- + 3 \; H_2O$$

Step 3   Balance electrically with electrons

$$2 \; I^- \longrightarrow I_2 + 2 \; e^-$$

$$6 \; H^+ + ClO_3^- + 6 \; e^- \longrightarrow Cl^- + 3 \; H_2O$$

Steps 4 and 5   Equalize the loss and gain of electrons; add half-reactions

$$3 \; (2 \; I^- \longrightarrow I_2 + 2 \; e^-)$$

$$\underline{6 \; H^+ + ClO_3^- + 6 \; e^- \longrightarrow Cl^- + 3 \; H_2O}$$

$$6 \; H^+ + ClO_3^- + 6 \; I^- \longrightarrow 3 \; I_2 + Cl^- + 3 \; H_2O$$

(f) $Cr_2O_7^{2-} + Fe^{2+} \longrightarrow Cr^{3+} + Fe^{3+}$

Step 1   Write half–reaction equations. Balance except H and O.

$$Fe^{2+} \longrightarrow Fe^{3+}$$

$$Cr_2O_7^{2-} \longrightarrow 2 \; Cr^{3+}$$

Step 2   Balance H and O using H_2O and H^+

$$Fe^{2+} \longrightarrow Fe^{3+}$$

$$14 \; H^+ + Cr_2O_7^{2-} \longrightarrow 2 \; Cr^{3+} + 7 \; H_2O$$

Step 3   Balance electrically with electrons

$$Fe^{2+} \longrightarrow Fe^{3+} + e^-$$

$$14 \; H^+ + Cr_2O_7^{2-} + 6 \; e^- \longrightarrow 2 \; Cr^{3+} + 7 \; H_2O$$

Steps 4 and 5   Equalize the loss and gain of electrons; add half-reactions

$$6 \; (Fe^{2+} \longrightarrow Fe^{3+} + e^-)$$

$$\underline{14 \; H^+ + Cr_2O_7^{2-} + 6 \; e^- \longrightarrow 2 \; Cr^{3+} + 7 \; H_2O}$$

$$14 \; H^+ + Cr_2O_7^{2-} + 6 \; Fe^{2+} \longrightarrow 2 \; Cr^{3+} + 6 \; Fe^{3+} + 7 \; H_2O$$

(g) $MnO_4^- + SO_2 \longrightarrow Mn^{2+} + SO_4^{2-}$

Step 1   Write half-reaction equations. Balance except H and O.

$$SO_2 \longrightarrow SO_4^{2-}$$
$$MnO_4^- \longrightarrow Mn^{2+}$$

Step 2   Balance H and O using H_2O and H^+

$$2\,H_2O + SO_2 \longrightarrow SO_4^{2-} + 4\,H^+$$
$$8\,H^+ + MnO_4^- \longrightarrow Mn^{2+} + 4\,H_2O$$

Step 3   Balance electrically with electrons

$$2\,H_2O + SO_2 \longrightarrow SO_4^{2-} + 4\,H^+ + 2\,e^-$$
$$8\,H^+ + MnO_4^- + 5\,e^- \longrightarrow Mn^{2+} + 4\,H_2O$$

Steps 4 and 5   Equalize the loss and gain of electrons; add half-reactions

$$5\,(2\,H_2O + SO_2 \longrightarrow SO_4^{2-} + 4\,H^+ + 2\,e^-)$$
$$\underline{2\,(8\,H^+ + MnO_4^- + 5\,e^- \longrightarrow Mn^{2+} + 4\,H_2O)}$$

$$2\,H_2O + 2\,MnO_4^- + 5\,SO_2 \longrightarrow 4\,H^+ + 2\,Mn^{2+} + 5\,SO_4^{2-}$$

(h) $H_3AsO_3 + MnO_4^- \longrightarrow H_3AsO_4 + Mn^{2+}$

Step 1   Write half-reaction equations. Balance except H and O.

$$H_3AsO_3 \longrightarrow H_3AsO_4$$
$$MnO_4^- \longrightarrow Mn^{2+}$$

Step 2   Balance H and O using H_2O and H^+

$$H_2O + H_3AsO_3 \longrightarrow 2\,H^+ + H_3AsO_4$$
$$8\,H^+ + MnO_4^- \longrightarrow Mn^{2+} + 4\,H_2O$$

Step 3   Balance electrically with electrons

$$H_2O + H_3AsO_3 \longrightarrow 2\,H^+ + H_3AsO_4 + 2\,e^-$$
$$8\,H^+ + MnO_4^- + 5\,e^- \longrightarrow Mn^{2+} + 4\,H_2O$$

Steps 4 and 5   Equalize the loss and gain of electrons; add half-reactions

$$5\,(H_2O + H_3AsO_3 \longrightarrow 2\,H^+ + H_3AsO_4 + 2\,e^-)$$
$$\underline{2\,(8\,H^+ + MnO_4^- + 5\,e^- \longrightarrow Mn^{2+} + 4\,H_2O)}$$

$$6\,H^+ + 5\,H_3AsO_4 + 2\,MnO_4^- \longrightarrow 5\,H_3AsO_4 + 2\,Mn^{2+} + 3\,H_2O$$

(i) $ClO_3^- + Cl^- \longrightarrow Cl_2$

Step 1   Write half reaction equations. Balance, except H and O.

$$Cl^- \longrightarrow Cl^0$$
$$ClO_3^- \longrightarrow Cl^0$$

Step 2   Balance H and O using H_2O and H^+

$$Cl^- \longrightarrow Cl^0$$

$$6\,H^+ + ClO_3^- \longrightarrow Cl^0 + 3\,H_2O$$

Step 3   Balance electrically with electrons

$$Cl^- \longrightarrow Cl^0 + e^-$$

$$6\,H^+ + ClO_3^- + 5\,e^- \longrightarrow Cl^0 + 3\,H_2O$$

Steps 4 and 5   Equalize the loss and gain of electrons; add half-reactions

$$5\,(Cl^- \longrightarrow Cl^0 + e^-)$$

$$\underline{6\,H^+ + ClO_3^- + 5\,e^- \longrightarrow Cl^0 + 3\,H_2O}$$

$$6\,H^+ + ClO_3^- + 5\,Cl^- \longrightarrow 3\,Cl_2 + 3\,H_2O$$

(j) $Cr_2O_7^{2-} + H_3AsO_3 \longrightarrow Cr^{3+} + H_3AsO_4$

Step 1   Write half-reaction equations. Balance, except H and O.

$$H_3AsO_3 \longrightarrow H_3AsO_4$$

$$Cr_2O_7^{2-} \longrightarrow 2\,Cr^{3+}$$

Step 2   Balance H and O using H_2O and H^+

$$H_2O + H_3AsO_3 \longrightarrow 2\,H^+ + H_3AsO_4$$

$$14\,H^+ + Cr_2O_7^{2-} \longrightarrow 2\,Cr^{3+} + 7\,H_2O$$

Step 3   Balance electrically with electrons

$$H_2O + H_3AsO_3 \longrightarrow 2\,H^+ + H_3AsO_4 + 2\,e^-$$

$$14\,H^+ + Cr_2O_7^{2-} + 6\,e^- \longrightarrow 2\,Cr^{3+} + 7\,H_2O$$

Steps 4 and 5   Equalize the loss and gain of electrons; add half-reactions

$$3\,(H_2O + H_3AsO_3 \longrightarrow 2\,H^+ + H_3AsO_4 + 2\,e^-)$$

$$\underline{14\,H^+ + Cr_2O_7^{2-} + 6\,e^- \longrightarrow 2\,Cr^{3+} + 7\,H_2O}$$

$$8\,H^+ + Cr_2O_7^{2-} + 6\,e^- \longrightarrow 2\,Cr^{3+} + 3\,H_3AsO_4 + 4\,H_2O$$

15. (a) $Cl_2 + IO_3^- \longrightarrow Cl^- + IO_4^-$

Step 1   Write half-reaction equations. Balance, except H and O.

$$IO_3^- \longrightarrow IO_4^-$$

$$Cl_2 \longrightarrow 2\,Cl^-$$

Step 2   Balance H and O using H_2O and H^+

$$H_2O + IO_3^- \longrightarrow IO_4^- + 2\,H^+$$

$$Cl_2 \longrightarrow 2\,Cl^-$$

Step 3 Add OH^- ions to both sides (same number as H^+ ions)

$$2\,OH^- + H_2O + IO_3^- \longrightarrow IO_4^- + 2\,H^+ + 2\,OH^-$$
$$Cl_2 \longrightarrow 2\,Cl^-$$

Step 4 Combine H^+ and OH^- to form H_2O; cancel H_2O where possible

$$2\,OH^- + H_2O + IO_3^- \longrightarrow IO_4^- + 2\,H_2O$$
$$Cl_2 \longrightarrow 2\,Cl^-$$
$$2\,OH^- + IO_3^- \longrightarrow IO_4^- + H_2O$$
$$Cl_2 \longrightarrow 2\,Cl^-$$

Step 5 Balance electrically with electrons

$$2\,OH^- + IO_3^- \longrightarrow IO_4^- + H_2O + 2\,e^-$$
$$Cl_2 + 2\,e^- \longrightarrow 2\,Cl^-$$

Step 6 Electron loss and gain is balanced

Step 7 Add half-reactions

$$2\,OH^- + IO_3^- + Cl_2 \longrightarrow IO_4^- + 2\,Cl^- + H_2O$$

(b) $\quad MnO_4^- + ClO_2^- \longrightarrow MnO_2 + ClO_4^-$

Step 1 Write half-reaction equations. Balance, except H and O.

$$ClO_2^- \longrightarrow ClO_4^-$$
$$MnO_4^- \longrightarrow MnO_2$$

Step 2 Balance H and O using H_2O and H^+

$$2\,H_2O + ClO_2^- \longrightarrow ClO_2^- + 4\,H^+$$
$$MnO_4^- + 4\,H^+ \longrightarrow MnO_2 + 2\,H_2O$$

Step 3 Add OH^- ions to both sides (same number as H^+ ions)

$$4\,OH^- + 2\,H_2O + ClO_2^- \longrightarrow ClO_4^- + 4\,H^+ + 4\,OH^-$$
$$4\,OH^- + MnO_4^- + 4\,H^+ \longrightarrow MnO_2 + 2\,H_2O + 4\,OH^-$$

Step 4 Combine H^+ and OH^- to form H_2O; cancel H_2O where possible

$$4\,OH^- + 2\,H_2O + ClO_2^- \longrightarrow ClO_4^- + 4\,H_2O$$
$$4\,H_2O + MnO_4^- \longrightarrow MnO_2 + 2\,H_2O + 4\,OH^-$$
$$4\,OH^- + ClO_2^- \longrightarrow ClO_4^- + 2\,H_2O$$
$$2\,H_2O + MnO_4^- \longrightarrow MnO_2 + 4\,OH^-$$

Step 5 Balance electrically with electrons

$$4\,OH^- + ClO_2^- \longrightarrow ClO_4^- + 2\,H_2O + 4\,e^-$$
$$2\,H_2O + MnO_4^- + 3\,e^- \longrightarrow MnO_2 + 4\,OH^-$$

Steps 6 and 7   Equalize gain and loss of electrons; add half–reactions

$$3 \; (4 \; OH^- + ClO_2^- \longrightarrow ClO_4^- + 2 \; H_2O + 4 \; e^-)$$

$$\underline{4 \; (2 \; H_2O + MnO_4^- + 3 \; e^- \longrightarrow MnO_2 + 4 \; OH^-)}$$

$$2 \; H_2O + 4 \; MnO_4^- + 3 \; ClO_2^- \longrightarrow 4 \; MnO_2 + 3 \; ClO_4^- + 4 \; OH^-$$

(c) $Se \longrightarrow SeO_3^{2-} + Se^{2-}$

Step 1   Write half–reaction equations. Balance, except H,O.

$$Se \longrightarrow SeO_3^{2-}$$

$$Se \longrightarrow Se^{2-}$$

Step 2   Balance H and O using H_2O and H^+

$$3 \; H_2O + Se \longrightarrow SeO_3^{2-} + 6 \; H^+$$

$$Se \longrightarrow Se^{2-}$$

Step 3   Add OH^- ions to both sides (same number as H^+ ions)

$$6 \; OH^- + 3 \; H_2O + Se \longrightarrow SeO_3^{2-} + 6 \; H^+ + 6 \; OH^-$$

$$Se \longrightarrow Se^{2-}$$

Step 4   Combine H^+ and OH^- to form H_2O; cancel H_2O where possible

$$6 \; OH^- + 3 \; H_2O + Se \longrightarrow SeO_3^{2-} + 6 \; H_2O$$

$$Se \longrightarrow Se^{2-}$$

$$6 \; OH^- + Se \longrightarrow SeO_3^{2-} + 3 \; H_2O$$

Step 5   Balance electrically with electrons

$$6 \; OH^- + Se \longrightarrow SeO_3^{2-} + 3 \; H_2O + 4 \; e^-$$

$$Se + 2 \; e^- \longrightarrow Se^{2-}$$

Steps 6 and 7   Equalize loss and gain of electrons; add half-reactions

$$6 \; OH^- + Se \longrightarrow SeO_3^{2-} + 3 \; H_2O + 4 \; e^-$$

$$\underline{2 \; (Se + 2 \; e^- \longrightarrow Se^{2-})}$$

$$6 \; OH^- + 3 \; Se \longrightarrow SeO_3^{2-} + 2 \; Se^{2-} + 3 \; H_2O$$

(d) $MnO_4^- + SO_3^{2-} \longrightarrow MnO_2 + SO_4^{2-}$

Step 1   Write half–reaction equations. Balance except H and O.

$$SO_3^{2-} \longrightarrow SO_4^{2-}$$

$$MnO_4^- \longrightarrow MnO_2$$

Step 2   Balance H and O using H_2O and H^+

$$H_2O + SO_3^{2-} \longrightarrow SO_4^{2-} + 2 \; H^+$$

$$MnO_4^- + 4 \; H^+ \longrightarrow MnO_2 + 2 \; H_2O$$

Step 3   Add OH^- ions to both sides (same number as H^+ ions)

$$2\,OH^- + H_2O + SO_3^{2-} \longrightarrow SO_4^{2-} + 2\,H^+ + 2\,OH^-$$

$$4\,OH^- + MnO_4^- + 4\,H^+ \longrightarrow MnO_2 + 3\,H_2O + 4\,OH^-$$

Step 4   Combine H^+ and OH^- to form H_2O; cancel H_2O where possible

$$2\,OH^- + H_2O + SO_3^{2-} \longrightarrow SO_4^{2-} + 2\,H_2O$$

$$MnO_4^- + 4\,H_2O \longrightarrow MnO_2 + 2\,H_2O + 4\,OH^-$$

$$2\,OH^- + SO_3^{2-} \longrightarrow SO_4^{2-} + H_2O$$

$$MnO_4^- + 2\,H_2O \longrightarrow MnO_2 + 4\,OH^-$$

Step 5   Balance electrically with electrons

$$2\,OH^- + SO_3^{2-} \longrightarrow SO_4^{2-} + H_2O + 2\,e^-$$

$$3\,e^- + MnO_4^- + 2\,H_2O \longrightarrow MnO_2 + 4\,OH^-$$

Steps 6 and 7   Equalize gain and loss of electrons; add half-reactions

$$3\,(2\,OH^- + SO_3^{2-} \longrightarrow SO_4^{2-} + H_2O + 2\,e^-)$$

$$\underline{2\,(MnO_4^- + 2\,H_2O + 3\,e^- \longrightarrow MnO_2 + 4\,OH^-)}$$

$$H_2O + 2\,MnO_4^- + 3\,SO_3^{2-} \longrightarrow 2\,MnO_2 + 3\,SO_4^{2-} + 2\,OH^-$$

(e) $ClO_2 + SbO_2^- \longrightarrow ClO_2^- + Sb(OH)_6^-$

Step 1   Write half-reaction equations. Balance, except H and O.

$$SbO_2^- \longrightarrow Sb(OH)_6^-$$

$$ClO_2 \longrightarrow ClO_2^-$$

Step 2   Balance H and O using H_2O and H^+

$$4\,H_2O + SbO_2^- \longrightarrow Sb(OH)_6^- + 2\,H^+$$

$$ClO_2 \longrightarrow ClO_2^-$$

Step 3   Add OH^- ions to both sides (same number as H^+ ions)

$$2\,OH^- + 4\,H_2O + SbO_2^- \longrightarrow Sb(OH)_6^- + 2\,H^+ + 2\,OH^-$$

$$ClO_2 \longrightarrow ClO_2^-$$

Step 4   Combine H^+ and OH^- to form H_2O; cancel H_2O where possible

$$2\,OH^- + 4\,H_2O + SbO_2^- \longrightarrow Sb(OH)_6^- + 2\,H_2O$$

$$ClO_2 \longrightarrow ClO_2^-$$

$$2\,OH^- + 2\,H_2O + SbO_2^- \longrightarrow Sb(OH)_6^-$$

Step 5   Balance electrically with electrons

$$2\,OH^- + 2\,H_2O + SbO_2^- \longrightarrow Sb(OH)_6^- + 2\,e^-$$

$$ClO_2 + e^- \longrightarrow ClO_2^-$$

Steps 6 and 7 Equalize gain and loss of electrons; add half-reactions

$$2 H_2O + 2 OH^- + SbO_2^- \longrightarrow Sb(OH)_6^- + 2 e^-$$

$$2 (ClO_2 + e^- \longrightarrow ClO_2^-)$$

$$2 H_2O + 2 ClO_2 + 2 OH^- + SbO_2^- \longrightarrow 2 ClO_2^- + Sb(OH)_6^-$$

(f) $Fe_3O_4 + MnO_4^- \longrightarrow Fe_2O_3 + MnO_2$

Step 1 Write half–reaction equations. Balance, except H and O.

$$2 Fe_3O_4 \longrightarrow 3 Fe_2O_3$$

$$MnO_4^- \longrightarrow MnO_2$$

Step 2 Balance H and O using H_2O and H^+

$$H_2O + 2 Fe_3O_4 \longrightarrow 3 Fe_2O_3 + 2 H^+$$

$$4 H^+ + MnO_4^- \longrightarrow MnO_2 + 2 H_2O$$

Step 3 Add OH^- ions to both sides (same number as H^+ ions)

$$2 OH^- + H_2O + 2 Fe_3O_4 \longrightarrow 3 Fe_2O_3 + 2 H^+ + 2 OH^-$$

$$4 OH^- + 4 H^+ + MnO_4^- \longrightarrow MnO_2 + 2 H_2O + 4 OH^-$$

Step 4 Combine H^+ and OH^- to form H_2O; cancel H_2O where possible

$$2 OH^- + H_2O + 2 Fe_3O_4 \longrightarrow 3 Fe_2O_3 + 2 H_2O$$

$$4 H_2O + MnO_4^- \longrightarrow MnO_2 + 2 H_2O + 4 OH^-$$

$$2 OH^- + 2 Fe_3O_4 \longrightarrow 3 Fe_2O_3 + H_2O$$

$$2 H_2O + MnO_4^- \longrightarrow MnO_2 + 4 OH^-$$

Step 5 Balance electrically with electrons

$$2 OH^- + 2 Fe_3O_4 \longrightarrow 3 Fe_2O_3 + H_2O + 2 e^-$$

$$2 H_2O + MnO_4^- + 3 e^- \longrightarrow MnO_2 + 4 OH^-$$

Steps 6 and 7 Equalize gain and loss of electrons; add half-reactions

$$3 (2 OH^- + 2 Fe_3O_4 \longrightarrow 3 Fe_2O_3 + H_2O + 2 e^-)$$

$$2 (2 H_2O + MnO_4^- + 3 e^- \longrightarrow MnO_2 + 4 OH^-)$$

$$H_2O + 6 Fe_3O_4 + 2 MnO_4^- \longrightarrow 9 Fe_2O_3 + 2 MnO_2 + 2 OH^-$$

(g) $BrO^- + Cr(OH)_4^- \longrightarrow Br^- + CrO_4^{2-}$

Step 1 Write half–reaction equations. Balance, except H, O

$$Cr(OH)_4^- \longrightarrow CrO_4^{2-}$$

$$BrO^- \longrightarrow Br^-$$

Step 2 Balance H and O using H_2O and H^+

$$Cr(OH)_4^- \longrightarrow CrO_4^{2-} + 4 H^+$$

$$2 \, H^+ + BrO^- \longrightarrow Br^- + H_2O$$

Step 3   Add OH^- ions to both sides (same number as H^+ ions)

$$4 \, OH^- + Cr(OH)_4^- \longrightarrow CrO_4^{2-} + 4 \, H^+ + 4 \, OH^-$$

$$2 \, OH^- + 2 \, H^+ + BrO^- \longrightarrow Br^- + H_2O + 2 \, OH^-$$

Step 4   Combine H^+ and OH^- to form H_2O; cancel H_2O where possible

$$4 \, OH^- + Cr(OH)_4^- \longrightarrow CrO_4^{2-} + 4 \, H_2O$$

$$2 \, H_2O + BrO^- \longrightarrow Br^- + H_2O + 2 \, OH^-$$

$$H_2O + BrO^- \longrightarrow Br^- + 2 \, OH^-$$

Step 5   Balance electrically with electrons

$$4 \, OH^- + Cr(OH)_4^- \longrightarrow CrO_4^{2-} + 4 \, H_2O + 3 \, e^-$$

$$H_2O + BrO^- + 2 \, e^- \longrightarrow Br^- + 2 \, OH^-$$

Steps 6 and 7   Equalize gain and loss of electrons; add half-reactions

$$2 \, (4 \, OH^- + Cr(OH)_4^- \longrightarrow CrO_4^{2-} + 4 \, H_2O + 3 \, e^-)$$

$$\underline{3 \, (H_2O + BrO^- + 2 \, e^- \longrightarrow Br^- + 2 \, OH^-)}$$

$$2 \, OH^- + 3 \, BrO^- + 2 \, Cr(OH)_4^- \longrightarrow 3 \, Br^- + 2 \, CrO_4^{2-} + 5 \, H_2O$$

(h) $P_4 \longrightarrow HPO_3^{2-} + PH_3$

Step 1   Write half–reaction equations. Balence except H and O.

$$P_4 \longrightarrow 4 \, HPO_3^{2-}$$

$$P_4 \longrightarrow 4 \, PH_3$$

Step 2   Balance H and O using H_2O and H^+

$$12 \, H_2O + P_4 \longrightarrow 4 \, HPO_3^{2-} + 20 \, H^+$$

$$12 \, H^+ + P_4 \longrightarrow 4 \, PH_3$$

Step 3   Add OH^- ions to both sides (same number as H^+ ions)

$$20 \, OH^- + 12 \, H_2O + P_4 \longrightarrow 4 \, HPO_3^{2-} + 20 \, H^+ + 20 \, OH^-$$

$$12 \, OH^- + 12 \, H^+ + P_4 \longrightarrow 4 \, PH_3 + 12 \, OH^-$$

Step 4   Combine H^+ and OH^- to form H_2O; cancel H_2O where possible

$$20 \, OH^- + 12 \, H_2O + P_4 \longrightarrow 4 \, HPO_3^{2-} + 20 \, H_2O$$

$$12 \, H_2O + P_4 \longrightarrow 4 \, PH_3 + 12 \, OH^-$$

$$20 \, OH^- + P_4 \longrightarrow 4 \, HPO_3^{2-} + 8 \, H_2O$$

Step 5   Balance electrically with electrons

$$20 \, OH^- + P_4 \longrightarrow 4 \, HPO_3^{2-} + 8 \, H_2O + 12 \, e^-$$

$$12 \, H_2O + P_4 + 12 \, e^- \longrightarrow 4 \, PH_3 + 12 \, OH^-$$

Steps 6 and 7 Loss and gain of electrons are equal; add half-reactions

$$4\,OH^- + 2\,H_2O + P_4 \longrightarrow 2\,HPO_3^{2-} + 2\,PH_3$$

(i) $Al + OH^- \longrightarrow Al(OH)_4^- + H_2$

Step 1   Write half–reaction equations. Balance, except H and O.

$$Al \longrightarrow Al(OH)_4^-$$
$$OH^- \longrightarrow H_2$$

Step 2   Balance H and O using H_2O and H^+

$$4\,H_2O + Al \longrightarrow Al(OH)_4^- + 4\,H^+$$
$$3\,H^+ + OH^- \longrightarrow H_2 + H_2O$$

Step 3   Add OH^- ions to both sides (same number as H^+ ions)

$$4\,OH^- + 4\,H_2O + Al \longrightarrow Al(OH)_4^- + 4\,H^+ + 4\,OH^-$$
$$3\,OH^- + 3\,H^+ + OH^- \longrightarrow H_2 + H_2O + 3\,OH^-$$

Step 4   Combine H^+ and OH^- to form H_2O; cancel H_2O where possible

$$4\,OH^- + 4\,H_2O + Al \longrightarrow Al(OH)_4^- + 4\,H_2O$$
$$3\,H_2O + OH^- \longrightarrow H_2 + H_2O + 3\,OH^-$$

$$4\,OH^- + Al \longrightarrow Al(OH)_4^-$$
$$2\,H_2O + OH^- \longrightarrow H_2 + 3\,OH^-$$

Step 5   Balance electrically with electrons

$$4\,OH^- + Al \longrightarrow Al(OH)_4^- + 3\,e^-$$
$$2\,H_2O + OH^- + 2\,e^- \longrightarrow H_2 + 3\,OH^-$$

Steps 6 and 7 Equalize gain and loss of electrons; add half-reactions

$$2\,(4\,OH^- + Al \longrightarrow Al(OH)_4^- + 3\,e^-)$$
$$3\,(2\,H_2O + OH^- + 2\,e^- \longrightarrow H_2 + 3\,OH^-)$$

$$2\,Al + 6\,H_2O + 2\,OH^- \longrightarrow 2\,Al(OH)_4^- + 3\,H_2$$

(j) $Al + NO_3^- \longrightarrow NH_3 + Al(OH)_4^-$

Step 1   Write half–reaction equation. Balance except H and O.

$$Al \longrightarrow Al(OH)_4^-$$
$$NO_3^- \longrightarrow NH_3$$

Step 2   Balance H and O using H_2O and H^+

$$4\,H_2O + Al \longrightarrow Al(OH)_4^- + 4\,H^+$$
$$9\,H^+ + NO_3^- \longrightarrow NH_3 + 3\,H_2O$$

Step 3   Add OH^- ions to both sides (same number as H^+ ions)

$$4\ OH^- + 4\ H_2O + Al \longrightarrow Al(OH)_4^- + 4\ H^+ + 4\ OH^-$$

$$9\ OH^- + 9\ H^+ + NO_3^- \longrightarrow NH_3 + 3\ H_2O + 9\ OH^-$$

Step 4   Combine H^+ and OH^- ions to form H_2O; cancel H_2O where possible

$$4\ OH^- + 4\ H_2O + Al \longrightarrow Al(OH)_4^- + 4\ H_2O$$

$$9\ H_2O + NO_3^- \longrightarrow NH_3 + 3\ H_2O + 9\ OH^-$$

$$4\ OH^- + Al \longrightarrow Al(OH)_4^-$$

$$6\ H_2O + NO_3^- \longrightarrow NH_3 + 9\ OH^-$$

Step 5   Balance electrically with electrons

$$4\ OH^- + Al \longrightarrow Al(OH)_4^- + 3\ e^-$$

$$6\ H_2O + NO_3^- + 8\ e^- \longrightarrow NH_3 + 9\ OH^-$$

Steps 6 and 7   Equalize gain and loss of electrons; add half-reactions

$$8\ (4\ OH^- + Al \longrightarrow Al(OH)_4^- + 3\ e^-)$$

$$\underline{3\ (6\ H_2O + NO_3^- + 8\ e^- \longrightarrow NH_3 + 9\ OH^-)}$$

$$8\ Al + 3\ NO_3^- + 18\ H_2O + 5\ OH^- \longrightarrow 3\ NH_3 + 8\ Al(OH)_4^-$$

16.   Oxidation and reduction are complementary processes because one does not occur without the other.

17. $Ca^{2+} + 2\ e^- \longrightarrow Ca$ cathode reaction, reduction

$2\ Br^- \longrightarrow Br_2 + 2\ e^-$ anode reaction, oxidation

18.   During electroplating of metals, the metal is plated by reducing the positive ions of the metal in the solution. The plating will occur at the cathode, the source of electrons.   With an alternating current, the polarity of the electrode would be constantly changing, so at one instant the metal would be plating and the next instant the metal would be dissolving.

19.   (a) $Pb + SO_4^{2-} \longrightarrow PbSO_4 + 2\ e^-$

$PbO_2 + SO_4^{2-} + 4\ H^+ + 2\ e^- \longrightarrow PbSO_4 + 2\ H_2O$

(b)   The first reaction is oxidation (Pb^0 is oxidized to Pb^{2+}).
The second reaction is reduction (Pb^{4+} is reduced to Pb^{2+})

(c)   The first reaction (oxidation) occurs at the anode of the battery.

20.   Since lead dioxide and lead(II) sulfate are insoluble, it is unnecessary to have salt bridges in the cells of a lead storage battery.

21.   The electolyte in a lead storage battery is dilute sulfuric acid.   In the discharge cycle, SO_4^{2-}, is removed from solution as it reacts with PbO_2 and H^+ to form $PbSO_4$ and H_2O. Therefore, the electrolyte solution contains less H_2SO_4 and becomes less dense.

22. If Hg^{2+} ions are reduced to metallic mercury, this would occur at the cathode, because reduction takes place at the cathode.

23. In both electrolytic and voltaic cells, oxidation and reduction reactions occur. In an electrolytic cell an electric current is forced through the cell causing a chemical change to occur. In voltaic cells, spontaneous chemical changes occur, generating an electric current.

24. In some voltaic cells, the reactants at the electrodes are in solution. For the cell to function, these reactants must be kept separated. A salt bridge permits movement of ions in the cell. This keeps the solution neutral with respect to the charged particles (ions) in the solution.

25. (a) The oxidizing agent is $KMnO_4$.

    (b) The reducing agent is HCl.

    (c) 5 moles of electrons $\quad 5\,e^- + Mn^{7+} \longrightarrow Mn^{2+}$

    $(5\text{ mol }e^-)\left(\dfrac{6.022 \times 10^{23}\,e^-}{\text{mol }e^-}\right) = 3.01 \times 10^{24}\ \dfrac{\text{electrons}}{\text{mol }KMnO_4}$

26. Iron will corrode (oxidize) in moist air. If the iron is in contact with copper and sea water (an electrolyte), the corrision rate increases dramatically. This is what occurred with the Staue of Liberty.

27. Patina is a blue-green product of the atmospheric corrosion of copper. It froms a coating which can protect the copper from further oxidation.

28. Photochromic glass is made from tetrahedrons of silicon and oxygen in a disorderly array with crystals of AgCl in between the tetrahedrons. Visible light passes through the glass but UV light triggers an oxidation-reduction reaction between Ag^+ and Cl^-. Cu^+ ions within the AgCl crystal react with the Cl^0 atoms to form Cu^{2+} and Cl^-, while the Ag^0 atoms move to the surface of the crystal and form clusters of Ag metal. When the glass is removed from the light, Cu^{2+} ions migrate to the surface and react with the Ag^0 reforming Ag^+ and Cu^+, cleaning the glass.

29. The correct statements are a, c, e, g, j, k, m, p, q, r, s

    (b) The oxidation number of molybdenum in Na_2MoO_4 is +6.

    (d) The process in which an atom or an ion loses electrons is called oxidation.

    (f) In the reaction $2\,Al + 3\,CuCl_2 \longrightarrow 2\,AlCl_3 + 3\,Cu$ aluminum is the reducing agent.

    (h) $Cu^0 \longrightarrow Cu^{2+} + 2\,e^-$ is a balanced oxidation half-reaction.

    (i) In the electrolysis of sodium chloride (brine) solution, Cl_2 gas is formed at the anode.

    (l) The reaction $Zn + MgCl_2 \longrightarrow Mg + ZnCl_2$ is not a spontaneous reaction.

    (n) Silver metal will not react with acids to liberate hydrogen gas.

    (o) Zinc is a better reducing agent than iron.

    (t) In an electrolytic cell, electrical energy is converted to chemical energy.

30. $3\,Ag + 4\,HNO_3 \longrightarrow 3\,AgNO_3 + NO + 2\,H_2O$

g Ag $\longrightarrow$ mol Ag $\longrightarrow$ mol NO

$(25.0 \text{ g Ag})\left(\dfrac{1 \text{ mol}}{107.9 \text{ g}}\right)\left(\dfrac{1 \text{ mol NO}}{3 \text{ mol Ag}}\right) = 0.0772 \text{ mol NO}$

31. $3\,Cl_2 + 6\,KOH \longrightarrow KClO_3 + 5\,KCl + 3\,H_2O$

mol $KClO_3$ $\longrightarrow$ mol Cl_2 $\longrightarrow$ L Cl_2

$(0.300 \text{ mol } KClO_3)\left(\dfrac{3 \text{ mol } Cl_2}{1 \text{ mol } KClO_3}\right)\left(\dfrac{22.4 \text{ L}}{1 \text{ mol}}\right) = 20.2 \text{ L } Cl_2$

32. $H_2O_2 + 2\,KMnO_4 + 3\,H_2SO_4 \longrightarrow 5\,O_2 + 2\,MnSO_4 + K_2SO_4 + 8\,H_2O$

mL H_2O_2 $\longrightarrow$ g H_2O_2 $\longrightarrow$ mol H_2O_2 $\longrightarrow$ mol $KMnO_4$ $\longrightarrow$ g $KMnO_4$

$(100. \text{ mL } H_2O_2)\left(\dfrac{1.031 \text{ g}}{mL}\right)(0.090)\left(\dfrac{1 \text{ mol}}{34.02 \text{ g}}\right)\left(\dfrac{2 \text{ mol } KMnO_4}{5 \text{ mol } H_2O_2}\right)\left(\dfrac{158.0 \text{ g}}{mol}\right) = 17 \text{ g } KMnO_4$

33. $Cr_2O_7{}^{2-} + 3\,H_3AsO_3 + 8\,H^+ \longrightarrow 2\,Cr^{3+} + 3\,H_3AsO_4 + 4\,H_2O$

g H_3AsO_3 $\longrightarrow$ mol H_3AsO_3 $\longrightarrow$ mol $Cr_2O_7{}^{2-}$ $\longrightarrow$ mL $Cr_2O_7{}^{2-}$

$(5.00 \text{ g } H_3AsO_4)\left(\dfrac{1 \text{ mol}}{125.9 \text{ g}}\right)\left(\dfrac{1 \text{ mol } Cr_2O_7{}^{2-}}{3 \text{ mol } H_3AsO_3}\right)\left(\dfrac{1000 \text{ mL}}{0.200 \text{ mol}}\right) = 66.2 \text{ mL of } 0.200 \text{ M } K_2Cr_2O_7 \text{ solution}$

34. $Cr_2O_7{}^{2-} + 6\,Fe^{2+} + 14\,H^+ \longrightarrow 2\,Cr^{3+} + 6\,Fe^{3+} + 7\,H_2O$

mL $FeSO_4$ $\longrightarrow$ mol $FeSO_4$ $\longrightarrow$ mol $Cr_2O_7{}^{2-}$ $\longrightarrow$ mL $Cr_2O_7{}^{2-}$

$(60.0 \text{ mL } FeSO_4)\left(\dfrac{0.200 \text{ mol}}{1000 \text{ mL}}\right)\left(\dfrac{1 \text{ mol } Cr_2O_7{}^{2-}}{6 \text{ mol } FeSO_4}\right)\left(\dfrac{1000 \text{ mL}}{0.200 \text{ mol}}\right) = 10.0 \text{ mL } 0.200 \text{ M } K_2Cr_2O_7 \text{ solution}$

35. $8\,KI + 5\,H_2SO_4 \longrightarrow 4\,I_2 + H_2S + 4\,K_2SO_4 + 4\,H_2O$

g I_2 $\longrightarrow$ mol I_2 $\longrightarrow$ mol KI $\longrightarrow$ g KI

$(2.79 \text{ g } I_2)\left(\dfrac{1 \text{ mol}}{253.8 \text{ g}}\right)\left(\dfrac{8 \text{ mol KI}}{2 \text{ mol } I_2}\right)\left(\dfrac{166.0 \text{ g}}{mol}\right) = 3.65 \text{ g KI in sample}$

$\left(\dfrac{3.65 \text{ g KI}}{4.00 \text{ g sample}}\right)(100) = 91.3\% \text{ KI}$

36. $3\,CuO + 2\,NH_3 \longrightarrow N_2 + 3\,Cu + 3\,H_2O$

L NH_3 $\longrightarrow$ mol NH_3 $\longrightarrow$ mol Cu $\longrightarrow$ g Cu

$(35.0 \text{ L } NH_3)\left(\dfrac{1 \text{ mol}}{22.4 \text{ L}}\right)\left(\dfrac{3 \text{ mol Cu}}{2 \text{ mol } NH_3}\right)\left(\dfrac{63.55 \text{ g}}{mol}\right) = 149 \text{ g Cu}$

37. $3\,Ag + 4\,HNO_3 \longrightarrow 3\,AgNO_3 + NO + 2\,H_2O$

mol Ag $\longrightarrow$ mol NO

$(0.500 \text{ mol Ag})\left(\dfrac{1 \text{ mol NO}}{3 \text{ mol Ag}}\right) = 0.167 \text{ mol NO}$

$PV = nRT \quad V = nRT/P$

$P = \left(\dfrac{744 \text{ torr}}{760 \text{ torr/atm}}\right) = 0.979 \text{ atm} \quad T = 301 \text{ K}$

$V = \dfrac{(0.167 \text{ mol NO})(0.0821 \text{ L atm/mol K})(301 \text{ K})}{(0.979 \text{ atm})} = 4.22 \text{ L NO}$

38. $2 \text{ Al} + 2 \text{ OH}^- + 6 \text{ H}_2\text{O} \longrightarrow 2 \text{ Al(OH)}_4^- + 3 \text{ H}_2$

g Al $\longrightarrow$ mol Al $\longrightarrow$ mol H_2

$(100. \text{ g Al})\left(\dfrac{1 \text{ mol Al}}{26.98 \text{ g}}\right)\left(\dfrac{3 \text{ mol H}_2}{2 \text{ mol Al}}\right) = 5.56 \text{ mol H}_2$

CHAPTER 19

NUCLEAR CHEMISTRY

1. (a)   Gamma radiation requires the most shielding.

   (b)   Alpha radiation requires the least shielding.

2. Alpha particles are deflected less than beta particles while passing through a magnetic field, because they are much heavier (more than 7,000 times heavier) than beta particles.

3. Pairs of nuclides that would be found in the fission reaction of U-235.   Any two nuclides, whose atomic numbers add up to 92 and mass numbers (in the range of 70-160) add up to 230-234. Examples include:

   $^{90}_{38}$Sr and $^{141}_{56}$Xe $^{139}_{56}$Ba and $^{94}_{36}$Kr $^{101}_{42}$Mo and $^{131}_{50}$Sn

4. Contributions to the early history of radioactivity include:

   (a)   Henri Becquerel: He discovered radioactivity.

   (b)   Marie and Pierre Curie: They discovered the elements polonium and radium.

   (c)   Wilhelm Roentgen: He discovered X rays and developed the technique of producing them.   While this was not a radioactive phenomenon, it triggered Becquerel's discovery of radioactivity.

   (d)   Earnest Rutherford: He discovered alpha and beta particles, established the link between radioactivity and transmutation, and produced the first successful man-made transmutation.

   (e)   Otto Hahn and Fritz Strassmann.   They were first to produce nuclear fission.

5. Chemical reactions are caused by atoms or ions coming together, so are greatly influenced by temperature and concentration, which affect the number of collisions.   Radioactivity is a spontaneous reaction of an individual nucleus, and is independent of such influences.

6. The term isotope is used with reference to atoms of the same element that contain different masses.   For example $^{12}_{6}$C and $^{14}_{6}$C.   The term nuclide is used in nuclear chemistry to infer any isotope of any atom.

7.

		Protons	Neutrons	Nucleons
(a)	$^{35}_{17}$Cl	17	18	35
(b)	$^{226}_{88}$Ra	88	138	226
(c)	$^{235}_{92}$U	92	143	235
(d)	$^{82}_{35}$Br	35	47	82

8. "Half-life" is the length of time required for one-half of a specific amount of a radionuclide to disintegrate. It is a convenient way of comparing stabilities of radioisotopes.

9. $(5 \times 10^9 \text{years})\left(\dfrac{1 \text{ half-life}}{7.6 \times 10^7 \text{years}}\right) = 66$ half-lives

Even if plutonium–224 had been present in large quantities five billion years ago, no measureable amount would survive after 66 half-lives.

10.

	charge	mass	nature of particles	penetrating power
Alpha	+2	4 amu	He nucleus	low
Beta	−1	$\dfrac{1}{1837}$ amu	electron	moderate
Gamma	0	0	electromagnetic radiation	high

11. (a) When a nucleus loses an alpha particle, its atomic number decreases by two, and its mass number decreases by four.

    (b) When a nucleus loses a beta particle, its atomic number increases by one, and its mass number remains unchanged.

12. Natural radioactivity is the spontaneous disintegration of those radioactive isotopes found in nature. Artificial radioactivity is the spontaneous disintegration of radioactive isotopes produced by man.

13. A disintegration series starting with a particular radionuclide, progressing stepwise by alpha and beta emissions to other radionuclides, ending at a stable nuclide. For example:

$$^{238}_{92}\text{U} \longrightarrow {}^{206}_{82}\text{Pb (stable)}$$

14. Transmutation is the conversion of one element into another by natural or artificial means. The nucleus of an atom is bombarded by various particles (alpha, beta, protons, etc.). The fast moving particles are captured by the nucleus, forming an unstable nucleus, which decays to another kind of atom. For example:

$$^{9}_{4}\text{Be} + {}^{4}_{2}\text{He} \longrightarrow {}^{12}_{6}\text{C} + {}^{1}_{0}\text{n}$$

15. Equations for alpha decays:

    (a) $^{218}_{85}\text{At} \longrightarrow {}^{4}_{2}\text{He} + {}^{214}_{83}\text{Bi}$

    (b) $^{221}_{87}\text{Fr} \longrightarrow {}^{4}_{2}\text{He} + {}^{217}_{85}\text{At}$

    (c) $^{192}_{78}\text{Pt} \longrightarrow {}^{4}_{2}\text{He} + {}^{188}_{76}\text{Os}$

    (d) $^{210}_{84}\text{Po} \longrightarrow {}^{4}_{2}\text{He} + {}^{206}_{82}\text{Pb}$

16. Equations for beta decay:

    (a) $^{14}_{6}\text{C} \longrightarrow {}^{0}_{-1}\text{e} + {}^{14}_{7}\text{N}$

    (b) $^{137}_{55}\text{Cs} \longrightarrow {}^{0}_{-1}\text{e} + {}^{137}_{56}\text{Ba}$

(c) $\quad ^{239}_{93}Np \longrightarrow ^{0}_{-1}e + ^{239}_{94}Pu$

(d) $\quad ^{90}_{38}Sr \longrightarrow ^{0}_{-1}e + ^{90}_{39}Y$

17. $\quad ^{239}_{90}Th \xrightarrow{-\alpha} ^{228}_{88}Ra \xrightarrow{-\beta} ^{228}_{89}Ac \xrightarrow{-\beta} ^{228}_{90}Th \xrightarrow{-\alpha} ^{224}_{88}Ra \xrightarrow{-\alpha} ^{220}_{86}Rn \xrightarrow{-\alpha} ^{216}_{84}Po \xrightarrow{-\alpha} ^{212}_{82}Pb$

$\quad \xrightarrow{-\beta} ^{212}_{83}Bi \xrightarrow{-\beta} ^{212}_{84}Po \xrightarrow{-\alpha} ^{208}_{82}Pb$

18. $\quad ^{237}_{93}Np$ loses seven alpha particles and four beta particles.

Determination of the final product: $^{209}_{83}Bi$

$\qquad$ nuclear charge $= 93 - 7(2) + 4(1) = 83$

$\qquad$ mass $= 237 - 7(4) = 209$

19. Decay of bismuth-211

$\quad ^{211}_{83}Bi \longrightarrow ^{4}_{2}He + ^{207}_{81}Tl \qquad ^{207}_{81}Tl \longrightarrow ^{0}_{-1}e + ^{207}_{82}Pb$

20.. (a) $\quad ^{13}_{6}C + ^{1}_{0}n \longrightarrow ^{14}_{6}C$

(b) $\quad ^{30}_{15}P \longrightarrow ^{30}_{14}S + ^{0}_{+1}e$

There are possibly other ways to produce each of these reactions.

21. (a) $\quad ^{27}_{13}Al + ^{4}_{2}He \longrightarrow ^{30}_{15}P + ^{1}_{0}n$

(b) $\quad ^{27}_{14}Si \longrightarrow ^{0}_{+1}e + ^{27}_{13}Al$

(c) $\quad ^{12}_{6}C + ^{2}_{1}H \longrightarrow ^{13}_{7}N + ^{1}_{0}n$

(d) $\quad ^{82}_{35}Br \longrightarrow ^{82}_{36}Kr + ^{0}_{-1}e$

(e) $\quad ^{66}_{29}Cu \longrightarrow ^{66}_{30}Zn + ^{0}_{-1}e$

(f) $\quad ^{0}_{-1}e + ^{7}_{4}Be \longrightarrow ^{7}_{3}Li$

(g) $\quad ^{27}_{13}Al + ^{4}_{2}He \longrightarrow ^{30}_{14}Si + ^{1}_{1}H$

(h) $\quad ^{85}_{37}Rb + ^{1}_{0}n \longrightarrow ^{82}_{35}Br + ^{4}_{2}He$

(i) $\quad ^{214}_{83}Bi \longrightarrow ^{4}_{2}He + ^{210}_{81}Tl$

22. Shorthand for nuclear equations

(a) $\quad ^{27}_{13}Al\,(\alpha, n)\,^{30}_{15}P$ $\qquad\qquad$ (c) $\quad ^{59}_{27}Co\,(n, \gamma)\,^{60}_{27}Co$

(b) $\quad ^{14}_{7}N\,(\alpha, p)\,^{17}_{8}O$ $\qquad\qquad$ (d) $\quad ^{249}_{98}Cf\,(^{12}_{6}C, 4n)\,^{257}_{104}Unq$

23. Nuclear equations from shorthand notation

(a) $\quad ^{23}_{11}Na + ^{1}_{1}H \longrightarrow ^{23}_{12}Mg + ^{1}_{0}n$

(b) $\quad ^{209}_{83}Bi + ^{2}_{1}H \longrightarrow ^{210}_{84}Po + ^{1}_{0}n$

(c) $\quad ^{238}_{92}U + ^{16}_{8}O \longrightarrow ^{249}_{100}Fm + 5\,^{1}_{0}n$

(d) $\quad ^{9}_{4}Be + ^{4}_{2}He \longrightarrow ^{12}_{6}C + ^{1}_{0}n$

24. The Geiger counter is used to detect the presence of radioactive material. The rays which are emitted enter the Geiger tube and ionize the argon gas present in the tube. The ions cause an electrical discharge in the tube, producing a current pulse, which is amplified. The pulse appears as a signal in the form of audible clicks and/or a flashing light, that can be measured with a calibrated meter.

25. Units of radioactivity

   (a) Curie (Ci).   A unit of radioactivity indicating the rate of decay. $1 \text{ Ci} = 3.7 \times 10^{10}$ disintegrations/second.

   (b) Roentgen.   A unit of exposure to gamma or X ray radiation.

   (c) Rad or Gray.   A unit of absorbed dose radiation.
        1 rad = 0.01 J absorbed per 1 kg of matter.
        1 Gy = 100 rad (tissue absorption)

   (d) Rem.   A unit of radiation dose equivalent.   The rem takes into account that the same dose from different sources of radiation does not produce the same degree of biological affect.

26. Two Germans, Otto Hahn and Fritz Strassmann, were the first scientists to report nuclear fission. The fission resulted from bombarding uranium nuclei with neutrons.

27. Natural uranium is 99+% U-238. Commercial nuclear reactors use U-235 enriched uranium as a fuel. Slow neutrons will cause the fission of U-235, but not U-238. Fast neutrons are capable of a nuclear reaction with U-238 to produce fissionable Pu-239. A breeder reactor converts nonfissionable U-238 to fissionable Pu-239, and in the process, manufactures more fuel than it consumes.

28. The fission reaction in a nuclear reactor and in an atomic bomb are essentially the same. The difference is that the fissioning is "wild" or uncontrolled in the bomb. In a nuclear reactor, the fissioning rate is controlled by means of moderators, such as graphite, to slow the neutrons and control rods of cadmium or boron to absorb some of the neutrons.

29. A certain amount of fissionable material (a critical mass) must be present before a self-supporting chain reaction can occur. Without a critical mass, too many neutrons from fissions will escape, and the reaction cannot reach a chain reaction status, unless at least one neutron is captured for every fission that occurs.

30. The terms "atomic bomb" and "hydrogen bomb" are not synonymous. The energy released by the explosion of an atomic bomb results from the fission of heavy atoms. The major part of the energy released by the explosion of a hydrogen bomb results from the fusion of light atoms.

31. The mass defect is the difference between the mass of an atom and the mass of the number of protons, neutrons, and electrons in that atom. The energy equivalent to this difference in mass is known as the nuclear binding energy.

32. When radioactive rays pass through normal matter, they cause that matter to become ionized (usually by knocking out electrons). Therefore, the radioactive rays are classified as ionizing radiation.

33. Some biological hazards associated with radioactivity are:

    (a) High levels of radiation can cause nausea, vomiting, diarrhea, and death. The radiation produces ionization in the cells, particularly in the nucleus of the cells.

    (b) Long-term exposure to low levels of radiation can weaken the body and cause malignant tumors.

    (c) Radiation can damage DNA molecules in the body causing mutations, which by reproduction, can be passed on to succeeding generations.

34. Strontium-90 has two characteristics that create concern. Its half-life is 28 years, so it remains active for a long period of time. The other characteristic is that Sr-90 is chemically similar to calcium, so that it is deposited in bone tissue along with calcium. Red blood cells are produced in the bone marrow. If the marrow is subjected to beta radiation from strontium -90, the red blood cells will be destroyed, increasing the incidence of leukemia and bone cancer.

35. A radioactive "tracer" is a radioactive material, whose presence is traced by a Geiger counter or some other detecting device. Tracers are often injected into the human body, animals, and plants to determine chemical pathways, rates of circulation, etc. For example, use of a tracer could determine the length of time for material to travel from the root sytem to the leaves in a tree.

36. In living species, the ratio of carbon-14 to carbon-12 is constant due to the constant C-14/C-12 ratio in the atmosphere and food sources. When a species dies, life processes stop. The C-14/C-12 ratio decreases with time because C-14 is radioactive and decays according to its half-life, while the amount of C-12 in the species remains constant. Thus, the age of an archaeological artifact containing carbon can be calculated by comparing the C-14/C-12 ratio in the artifact with the C-14/C-12 ratio in the living species.

37. Radioactivity could be used to locate a leak in an underground pipe by using a water soluble tracer element. Dissolve the tracer in water and pass the water through the pipe. Test the ground along the path of the pipe with a Geiger counter until radioactivity from the leak is detected. Dig.

38. The half-life of carbon-14 is 5668 years.

$$(4 \ \times 10^6 \text{ years})\left(\frac{1 \text{ half-life}}{5668 \text{ years}}\right) \ = \ 7 \times 10^2 \text{ half-lives}$$

700 half-lives would pass in 4 million years. Not enough C-14 would remain to allow detection with any degree of reliability. Such dating would not prove useful.

39. $(112 \text{ years})\left(\frac{1 \text{ half-life}}{28 \text{ years}}\right) \ = \ 4 \text{ half-lives}$

In 4 half-lives 1/16th $(1/2)^4$ of the starting amount would remain.

$\frac{1.00 \text{ mg Sr-90}}{16} \ = \ 0.0625 \text{ mg Sr-90 remains after 112 years.}$

40. $(0.0100 \text{ g RaCl}_2)\left(\frac{226.0 \text{ g Ra}}{296.9 \text{ g RaCl}_2}\right)\left(\frac{\$50,000}{1 \text{ g Ra}}\right) = \$381$

41. $\frac{240}{2} = 120$; $\frac{120}{2} = 60$; $\frac{60}{2} = 30$; 3 half-lives are required

to reduce the count from 240 to 30 counts/min.

$1980 + (3 \times 28) = 2064$ A.D. One eighth of the original amount remains.

42. 100% to 25% requires 2 half-lives. The half-life of C-14 is 5668 years. The specimen will be the age of two half-lives:

$(2)(5668 \text{ years}) = 1.1 \times 10^4$ years old.

43. 16.0 g $\longrightarrow$ 8.0 g $\longrightarrow$ 4.0 g $\longrightarrow$ 2.0 g $\longrightarrow$ 1.0 g $\longrightarrow$ 0.50 g

16.0 g to 0.50 g requires five half-lives.

$\frac{90 \text{ minutes}}{5 \text{ half-lives}} = 18$ minutes/half-life

44. (a) Calculated mass

3 protons	$3(1.0073 \text{ g})$	$= 3.0219$ g
4 neutrons	$4(1.0087 \text{ g})$	$= 4.0348$ g
3 electrons	$3(0.00055 \text{ g})$	$= 0.0017$ g
calculated mass		7.0584 g

Mass defect = calculated mass - actual mass

Mass defect 7.0584 g $-$ 7.0160 g $= 0.0424$ g/mol

(b) Binding energy

$\left(\frac{0.0424 \text{ g}}{\text{mol}}\right)\left(\frac{9.0 \times 10^{13} \text{ J}}{\text{g}}\right) = 3.8 \times 10^{12}$ J/mol

45. (a) $^{235}_{92}\text{U} + ^{1}_{0}\text{n} \longrightarrow ^{94}_{38}\text{Sr} + ^{139}_{54}\text{Xe} + 3\,^{1}_{0}\text{n} + $ energy

Mass loss = mass of reactants $-$ mass of products

Mass of reactants $= 235.0439 + 1.0087 = 236.0526$

Mass of products $= 93.9154 + 138.9179 + 3(1.0087) = 235.8594$

Mass lost $= 236.0526$ amu $- 235.8594$ amu $= 0.1932$ amu

$(0.1932 \text{ amu})\left(\frac{1.000 \text{ g}}{6.022 \times 10^{23} \text{ amu}}\right)\left(\frac{9.0 \times 10^{13} \text{ J}}{1.00 \text{ g}}\right) = 2.9 \times 10^{-11}$ J/atom U-235

(b) $\left(\frac{2.9 \times 10^{-11} \text{ J}}{\text{atom}}\right)\left(\frac{6.022 \times 10^{23} \text{ atoms}}{\text{mol}}\right) = 1.7 \times 10^{13}$ J/mol

(c) $\left(\frac{0.1932 \text{ amu}}{236.0526 \text{ amu}}\right)(100) = 0.08185\%$ mass loss

46. (a) $^{1}_{1}\text{H} + ^{2}_{1}\text{H} \longrightarrow ^{3}_{2}\text{He} + $ energy

Mass loss = mass reactants - mass products

Mass reactants $= 1.00782 + 2.01410 = 3.02192$ g/mol

Mass products $=$ 3.01603 g/mol

Mass loss $=$ 3.022192 $-$ 3.01603 $=$ 0.00589 g/mol

$$\left(\frac{0.00589 \text{ g}}{\text{mol}}\right)\left(\frac{9.0 \text{ x } 10^{13} \text{ J}}{\text{g}}\right) = 5.3 \text{ x } 10^{11} \text{ J/mol}$$

(b) $\left(\frac{0.00589 \text{ g}}{3.02192 \text{ g}}\right)(100) = 0.195\%$ mass loss

47. $^{235}_{92}\text{U} \longrightarrow {}^{207}_{82}\text{Pb}$

Mass loss: 235 - 207 $=$ 28

Net proton loss (atomic number):   92 p $-$ 82 p $=$ 10 p

The mass loss is equivalent to 7 alpha particles (28/4).   A loss of 7 alpha particles gives a loss of 14 protons.   A decrease in the atomic number to 78 (14 protons) is due to the loss of 7 alpha particles, $(92 - 14 = 78)$.   Therefore, a loss of 4 beta particles is required to increase the atomic number from 78 to 82.

The total loss $=$ 7 alpha particles and 4 beta particles.

48.   (a)   Geiger counter - radiation passes through a thin glass window into a chamber filled with argon gas and containing two electrodes.   Some of the argon ionizes, sending a momentary electrical impulse between the electrodes to the detector.   This signal is amplified electronically and read out on a counter or as a series of clicks.

(b)   Scintillation counter - radiation strikes a scintillator, which is composed of molecules that emits light in the presence of ionizing radiation.   A light sensitive detector counts the flashes and converts them into a digital readout.

(c)   Film badge - radiation penetrates a film holder.   The silver grains in the film darken when exposed to radiation.   The film is developed at regular intervals.

49.   (3 days)(24 hours/day) $=$ 72 hours

72 hr $+$ 6 hr $=$ 78 hr

$\dfrac{78 \text{ hr}}{13 \frac{\text{hr}}{t_{0.5}}} = 6$ half-lives $(10 \text{ mg})\left(\frac{1}{2}\right)^6 = 0.16$ mg remaining

50.   Fission is the process of splitting a large nucleus into two roughly equal mass pieces.   Fission occurs in nuclear reactors, or atomic bombs.

Fusion is the process of combining two relatively small nuclei to form a single larger nucleus. Fusion occurs on the sun, or in a hydrogen bomb.

51.   The correct statements are c, e, f, g, j, k, n, o, q, r, t

(a)   Radioactivity was discovered by Henri Becquerel.

(b)   An atom of $^{59}_{28}\text{Ni}$ has 28 protons.

(d)   The emission of an alpha particle from the nucleus of an atom lowers its atomic number by 2 and lowers its mass number by 4.

(h) The gamma ray has the greatest penetrating power of all the rays emitted from the nucleus of an atom.

(i) Radioactivity is due to an unstable ratio of neutrons to protons in an atom.

(l) The disintegration of $^{226}_{88}\text{Ra}$ into $^{214}_{83}\text{Po}$ involves the loss of 3 alpha particles and one beta particle.

(m) If 1.0 g of a radionuclide has a half-life of 7.2 days, the half-life of 0.50 g of that nuclide is 7.2 days.

(p) Radiocarbon dating of archaeological artifacts is based on a decrease in the C-14/C-12 ratio in the object.

(s) Carbon-14 is produced in the atmosphere.

CHAPTER 20

INTRODUCTION TO ORGANIC CHEMISTRY

1. Carbon atoms have the characteristic of bonding extensively with one another. They form organic compounds containing carbon chains of varying lengths and structure. Consequently, a great many compounds of carbon exist.

2. The most common geometric pattern of covalent bonds about carbon atoms is the tetrahedral arrangement of bonds. A simple example is the methane molecule, CH_4, with hydrogen atoms at the corners of the tetrahedron and the carbon atom at the center.

3. In addition to single bonds, carbon atoms can also form double and triple bonds.

4. Lewis structures for:

   (a)   a carbon atom

   $\cdot \overset{\displaystyle .}{\underset{\displaystyle .}{C}} \cdot$

   (b)   molecules of

   $$H : \overset{\displaystyle H}{\underset{\displaystyle H}{\ddot{C}}} : H \qquad H : \overset{\displaystyle H}{\underset{\displaystyle H}{\ddot{C}}} :: \overset{\displaystyle H}{\underset{\displaystyle H}{\ddot{C}}} : H \qquad H : C ::: C : H$$

   methane   ethylene   acetylene

5. Names and formulas of the first ten normal alkanes

   (a)   methane

   $$H - \overset{\displaystyle H}{\underset{\displaystyle H}{C}} - H$$

   (b)   ethane

   $$H - \overset{\displaystyle H}{\underset{\displaystyle H}{C}} - \overset{\displaystyle H}{\underset{\displaystyle H}{C}} - H$$

   (c)   propane

   $$H - \overset{\displaystyle H}{\underset{\displaystyle H}{C}} - \overset{\displaystyle H}{\underset{\displaystyle H}{C}} - \overset{\displaystyle H}{\underset{\displaystyle H}{C}} - H$$

(d)   butane

$$H - \underset{\overset{|}{H}}{\overset{\overset{H}{|}}{C}} - \underset{\overset{|}{H}}{\overset{\overset{H}{|}}{C}} - \underset{\overset{|}{H}}{\overset{\overset{H}{|}}{C}} - \underset{\overset{|}{H}}{\overset{\overset{H}{|}}{C}} - H$$

(e)   pentane

$$H - \underset{\overset{|}{H}}{\overset{\overset{H}{|}}{C}} - \underset{\overset{|}{H}}{\overset{\overset{H}{|}}{C}} - \underset{\overset{|}{H}}{\overset{\overset{H}{|}}{C}} - \underset{\overset{|}{H}}{\overset{\overset{H}{|}}{C}} - \underset{\overset{|}{H}}{\overset{\overset{H}{|}}{C}} - H$$

(f)   hexane

$$H - \underset{\overset{|}{H}}{\overset{\overset{H}{|}}{C}} - \underset{\overset{|}{H}}{\overset{\overset{H}{|}}{C}} - \underset{\overset{|}{H}}{\overset{\overset{H}{|}}{C}} - \underset{\overset{|}{H}}{\overset{\overset{H}{|}}{C}} - \underset{\overset{|}{H}}{\overset{\overset{H}{|}}{C}} - \underset{\overset{|}{H}}{\overset{\overset{H}{|}}{C}} - H$$

(g)   heptane

$$H - \underset{\overset{|}{H}}{\overset{\overset{H}{|}}{C}} - \underset{\overset{|}{H}}{\overset{\overset{H}{|}}{C}} - \underset{\overset{|}{H}}{\overset{\overset{H}{|}}{C}} - \underset{\overset{|}{H}}{\overset{\overset{H}{|}}{C}} - \underset{\overset{|}{H}}{\overset{\overset{H}{|}}{C}} - \underset{\overset{|}{H}}{\overset{\overset{H}{|}}{C}} - \underset{\overset{|}{H}}{\overset{\overset{H}{|}}{C}} - H$$

(h)   octane

$$H - \underset{\overset{|}{H}}{\overset{\overset{H}{|}}{C}} - \underset{\overset{|}{H}}{\overset{\overset{H}{|}}{C}} - \underset{\overset{|}{H}}{\overset{\overset{H}{|}}{C}} - \underset{\overset{|}{H}}{\overset{\overset{H}{|}}{C}} - \underset{\overset{|}{H}}{\overset{\overset{H}{|}}{C}} - \underset{\overset{|}{H}}{\overset{\overset{H}{|}}{C}} - \underset{\overset{|}{H}}{\overset{\overset{H}{|}}{C}} - \underset{\overset{|}{H}}{\overset{\overset{H}{|}}{C}} - H$$

(i)   nonane

$$H - \underset{\overset{|}{H}}{\overset{\overset{H}{|}}{C}} - \underset{\overset{|}{H}}{\overset{\overset{H}{|}}{C}} - \underset{\overset{|}{H}}{\overset{\overset{H}{|}}{C}} - \underset{\overset{|}{H}}{\overset{\overset{H}{|}}{C}} - \underset{\overset{|}{H}}{\overset{\overset{H}{|}}{C}} - \underset{\overset{|}{H}}{\overset{\overset{H}{|}}{C}} - \underset{\overset{|}{H}}{\overset{\overset{H}{|}}{C}} - \underset{\overset{|}{H}}{\overset{\overset{H}{|}}{C}} - \underset{\overset{|}{H}}{\overset{\overset{H}{|}}{C}} - H$$

(J)   decane

$$H - \underset{\overset{|}{H}}{\overset{\overset{H}{|}}{C}} - \underset{\overset{|}{H}}{\overset{\overset{H}{|}}{C}} - \underset{\overset{|}{H}}{\overset{\overset{H}{|}}{C}} - \underset{\overset{|}{H}}{\overset{\overset{H}{|}}{C}} - \underset{\overset{|}{H}}{\overset{\overset{H}{|}}{C}} - \underset{\overset{|}{H}}{\overset{\overset{H}{|}}{C}} - \underset{\overset{|}{H}}{\overset{\overset{H}{|}}{C}} - \underset{\overset{|}{H}}{\overset{\overset{H}{|}}{C}} - \underset{\overset{|}{H}}{\overset{\overset{H}{|}}{C}} - \underset{\overset{|}{H}}{\overset{\overset{H}{|}}{C}} - H$$

6. Alkyl groups

(a)   methyl

$$H - \underset{\overset{|}{H}}{\overset{\overset{H}{|}}{C}} - \qquad CH_3 -$$

(b)   ethyl

$$H - \underset{\overset{|}{H}}{\overset{\overset{H}{|}}{C}} - \underset{\overset{|}{H}}{\overset{\overset{H}{|}}{C}} - \qquad CH_3CH_2 -$$

(c)   propyl

$$H - \underset{\overset{|}{H}}{\overset{\overset{H}{|}}{C}} - \underset{\overset{|}{H}}{\overset{\overset{H}{|}}{C}} - \underset{\overset{|}{H}}{\overset{\overset{H}{|}}{C}} - \qquad CH_3CH_2CH_2 -$$

(d)   isopropyl

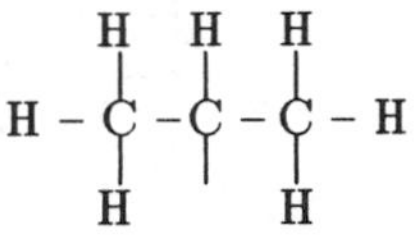

$$H - \overset{\overset{\displaystyle H}{|}}{\underset{\underset{\displaystyle H}{|}}{C}} - \overset{\overset{\displaystyle H}{|}}{\underset{}{C}} - \overset{\overset{\displaystyle H}{|}}{\underset{\underset{\displaystyle H}{|}}{C}} - H \qquad (CH_3)_2CH -$$

(e)   n–butyl

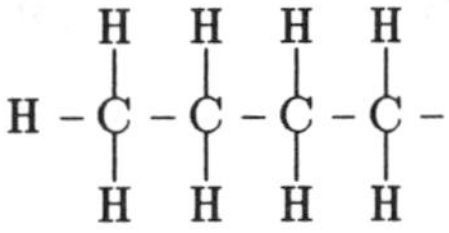

$$H - \overset{\overset{\displaystyle H}{|}}{\underset{\underset{\displaystyle H}{|}}{C}} - \overset{\overset{\displaystyle H}{|}}{\underset{\underset{\displaystyle H}{|}}{C}} - \overset{\overset{\displaystyle H}{|}}{\underset{\underset{\displaystyle H}{|}}{C}} - \overset{\overset{\displaystyle H}{|}}{\underset{\underset{\displaystyle H}{|}}{C}} - \qquad CH_3\,CH_2CH_2CH_2 -$$

(f)   s–butyl

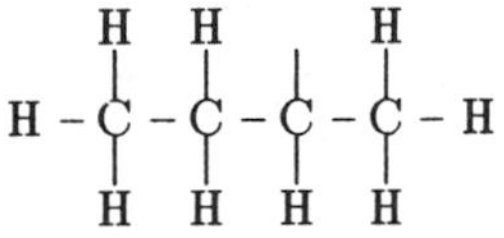

$$H - \overset{\overset{\displaystyle H}{|}}{\underset{\underset{\displaystyle H}{|}}{C}} - \overset{\overset{\displaystyle H}{|}}{\underset{\underset{\displaystyle H}{|}}{C}} - \overset{}{\underset{\underset{\displaystyle H}{|}}{C}} - \overset{\overset{\displaystyle H}{|}}{\underset{\underset{\displaystyle H}{|}}{C}} - H \quad CH_3CH_2CHCH_3$$

(g)   isobutyl

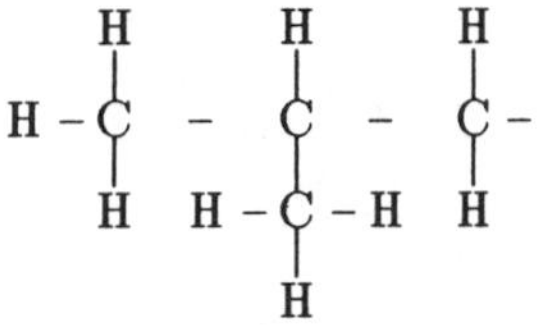

$$(CH_3)_2\ CHCH_2 -$$

(h)   t–butyl

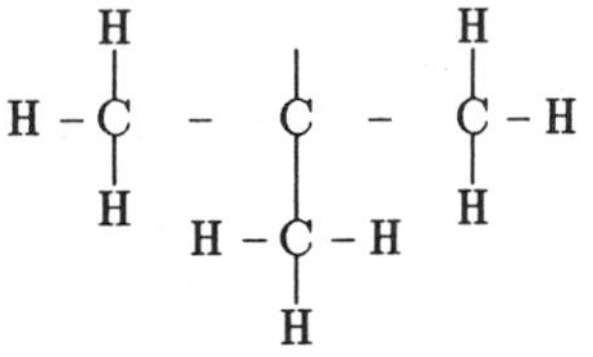

$$(CH_3)_3C -$$

7. Names of alkyl groups

   (a) $C_5H_{11} -$ pentyl

   (b) $C_7H_{15} -$ heptyl

   (c) $C_8H_{17} -$ octyl

   (d) $C_{10}H_{21} -$ decyl

8. Natural gas, petroleum, and coal are the three principle sources of hydrocarbons.

9. Compound (a) is an alkene and belongs to a homologous series of compounds represented by the formula, C_nH_{2n}.

   Compounds (b), (c), and (d) are all alkanes, C_nH_{2n+2}

10. The word ethylene represents a compound containing a double bond.   All other words represent structures having no double bonds.

11. **(a)** Hexane

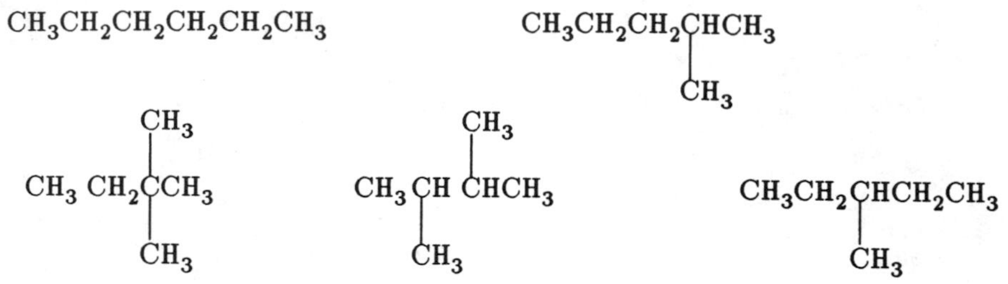

$CH_3CH_2CH_2CH_2CH_2CH_3$ $\qquad\qquad$ $CH_3CH_2CH_2\underset{\underset{CH_3}{|}}{C}HCH_3$

$CH_3\,CH_2\underset{\underset{CH_3}{|}}{\overset{\overset{CH_3}{|}}{C}}CH_3$ $\qquad$ $CH_3\,\underset{\underset{CH_3}{|}}{C}H\,\overset{\overset{CH_3}{|}}{C}HCH_3$ $\qquad$ $CH_3CH_2\underset{\underset{CH_3}{|}}{C}HCH_2CH_3$

(b) Heptane

$CH_3CH_2CH_2CH_2CH_2CH_2CH_3$ $\qquad\qquad$ $CH_3CH_2CH_2\underset{\underset{CH_3}{|}}{C}H_2CHCH_3$

$CH_3CH_2\underset{\underset{CH_3}{|}}{\overset{\overset{CH_3}{|}}{C}}CH_2CH_3$ $\qquad\qquad$ $CH_3CH_2CH_2\underset{\underset{CH_3}{|}}{\overset{\overset{CH_3}{|}}{C}}CH_3$

$CH_3CH_2CH_2\,\underset{\underset{CH_3}{|}}{C}HCH_2CH_3$ $\qquad\qquad$ $CH_3\,\underset{\underset{CH_3}{|}}{C}HCH_2\,\underset{\underset{CH_3}{|}}{C}HCH_3$

$CH_3CH_2\underset{\underset{CH_3}{|}}{C}H\,\overset{\overset{CH_3}{|}}{C}HCH_3$ $\qquad\qquad$ $CH_3\,\underset{\underset{CH_3}{|}}{C}H-\underset{\underset{CH_3}{|}}{\overset{\overset{CH_3}{|}}{C}}CH_3$

$CH_3CH_2\underset{\underset{CH_2CH_3}{|}}{C}HCH_2CH_3$

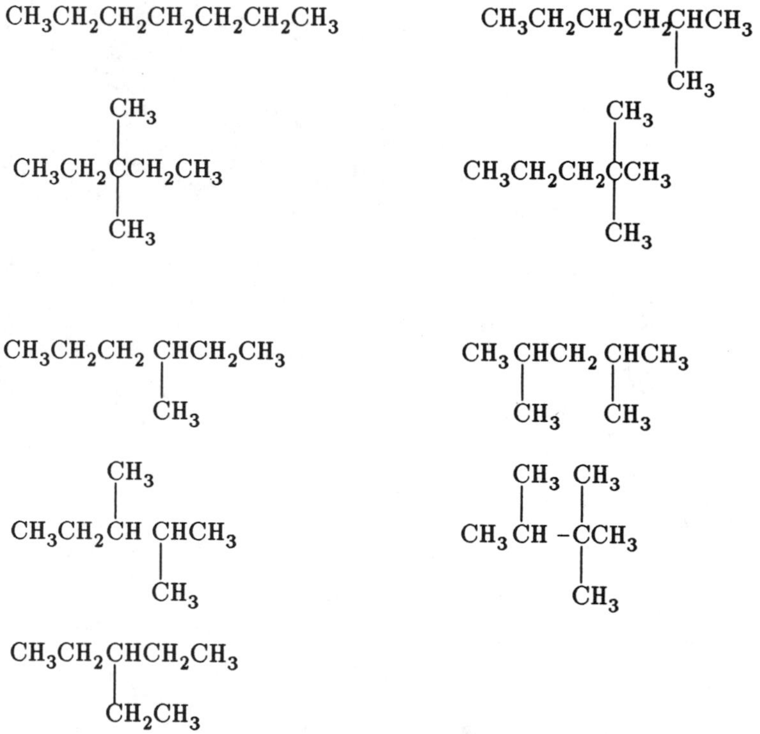

12. Common and IUPAC names

(a) CH_3CH_2Cl $\qquad$ ethyl chloride; chloroethane

(b) $CH_3CHClCH_3$ $\qquad$ isopropyl chloride;
2–chloropropane

(c) $(CH_3)_2CHCH_2Cl$ $\qquad$ isobutyl chloride;
1–chloro–2–methylpropane

(d) $CH_3CH_2CH_2Cl$ $\qquad$ *n*–propyl chloride; 1–chloropropane

(e) $(CH_3)_3CCl$ *t*–butyl chloride;
2 chloro–2–methylpropane

(f) $CH_3CHClCH_2CH_3$ *sec*–butyl chloride; 2–chlorobutane

13. (a) 4–ethyl–2–methylheptane

(b) 4,5–diethyl–2–methylheptane

(c) 3,6–dimethyloctane

(d) 3–ethyl–2–methylhexane (3–isopropylhexane)

14. Structural formulas

(a) 2,4–dimethylpentane $(CH_3)_2CHCH_2CH(CH_3)_2$

(b) 2,2–dimethylpentane $(CH_3)_3CCH_2CH_2CH_3$

(c) 3–isopropyloctane $CH_3CH_2\ CHCH_2CH_2CH_2CH_2CH_3$
$\underset{\displaystyle CH(CH_3)_2}{|}$

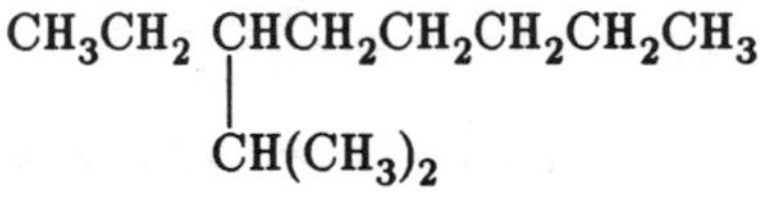

(d) 4–ethyl–2–methylhexane $CH_3CH_2\ CHCH_2CH(CH_3)_2$
$\underset{\displaystyle CH_2CH_3}{|}$

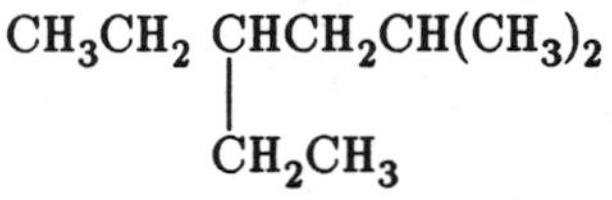

(e) 4–*t*–butylheptane $CH_3CH_2CH_2\ CHCH_2CH_2CH_3$
$\underset{\displaystyle C(CH_3)_3}{|}$

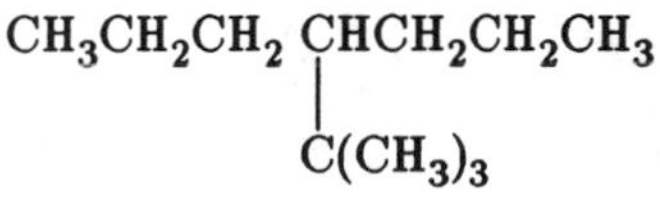

(f) 4–ethyl–7–isopropyl–
2,4,8–trimethyldecane

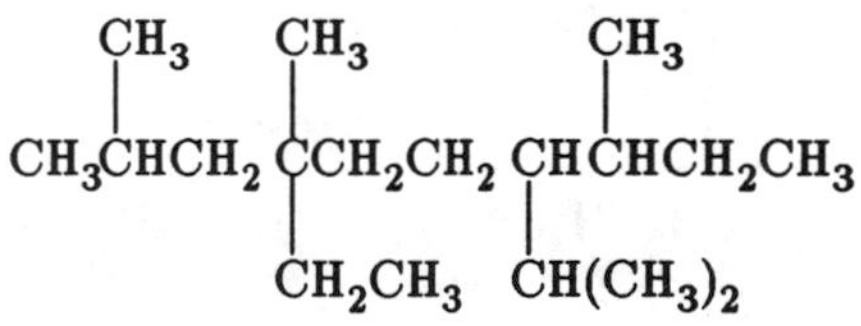

$$\overset{\displaystyle CH_3\ \ \ CH_3\ \ \ \ \ \ \ \ CH_3}{CH_3CHCH_2\ CCH_2CH_2\ CHCHCH_2CH_3}$$
$$\underset{\displaystyle CH_2CH_3\ \ \ CH(CH_3)_2}{|\ \ \ \ \ \ \ \ \ \ \ \ |}$$

15. (a) $CH_3CHCH_2CH_3$
$\underset{\displaystyle CH_3}{|}$

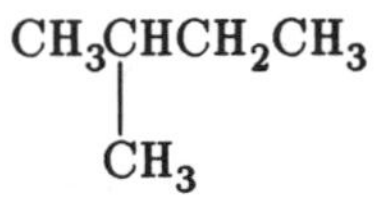

3–methylbutane is an incorrect name because the carbon atoms were numbered from the wrong end of the chain. The correct name is 2–methylbutane.

(b) $CH_3CHCH_2CH_3$
$\underset{\displaystyle CH_2CH_3}{|}$

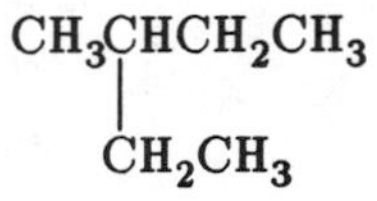

2–ethylbutane is incorrect. The longest carbon chain in the molecule was not used in determining the root name. The correct name is 3–methylpentane.

(c)

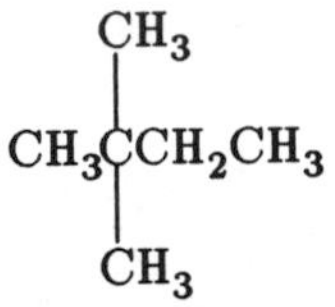

$$CH_3\underset{\underset{CH_3}{|}}{\overset{\overset{CH_3}{|}}{C}}CH_2CH_3$$

2–dimethylbutane is incorrect. Each methyl group is attached to carbon 2 on the chain and requires a number to identify its location on the chain. The correct name is 2,2–dimethyl pentane.

(d)

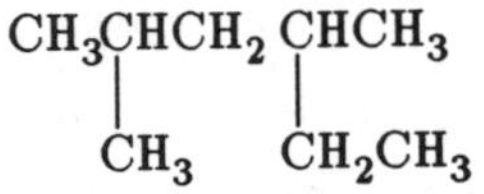

$$CH_3\underset{\underset{CH_3}{|}}{C}HCH_2\underset{\underset{CH_2CH_3}{|}}{C}HCH_3$$

2–ethyl–4–methylpentane is incorrect. The longest continuous chain in the molecule was not used to name the compound. The correct name is 2,4–dimethylhexane.

16. The single most important reaction of alkanes is combustion.

17. The sample had to be a mixture of two butanes, n–butane and isobutane, because each leads to only two monobromo compounds.

n–butane yields $CH_3CH_2CH_2CH_2Br$ and $CH_3CH_2CHBrCH_3$

isobutane yields $(CH_3)_2CHCH_2Br$ and $(CH_3)_3CBr$

18. (a)   (1) CH_3Br

(b)   (1) CH_2Cl_2

(c)   (1) C_2H_5Br

(d)   (2) $CH_3CH_2CH_2Cl$ $CH_3\underset{\underset{Cl}{|}}{C}HCH_3$

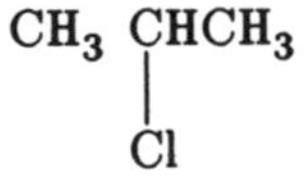

(e)   (4) $CH_3CH_2CH_2CH_2I$ $CH_3\underset{\underset{CH_2CH_3}{|}}{C}HI$

$$CH_3\underset{\underset{CH_3}{|}}{\overset{\overset{CH_3}{|}}{C}}-I$$ $CH_3\underset{\underset{CH_3}{|}}{C}HCH_2I$

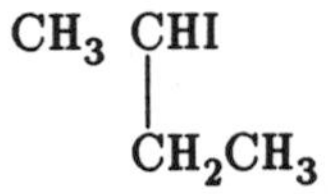

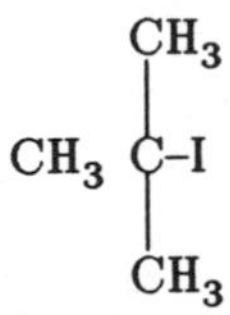

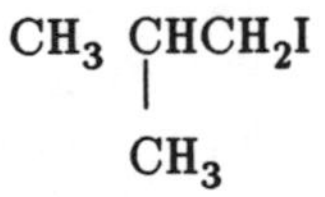

(f)   (4) $CH_3CH_2CHCl_2$ $CH_3CCl_2CH_3$

              $CH_3CHClCH_2Cl$ $CH_2ClCH_2CH_2Cl$

(g)   (5) $CH_3CH_2CHBrCl$ $CH_3CHClCH_2Br$

              $CH_3CHBrCH_2Cl$ $CH_2ClCH_2CH_2Br$

              $CH_3CClBrCH_3$

(h)   (9) $CH_3CH_2CH_2CHCl_2$ $CH_3CH_2CHClCH_2Cl$

              $CH_3CHClCH_2CH_2Cl$ $CH_2ClCH_2CH_2CH_2Cl$

              $CH_3CH_2CCl_2CH_3$ $CH_3CHClCHClCH_3$

$$CH_3\underset{\underset{\displaystyle CH_3}{|}}{CH}CHCl_2 \qquad\qquad CH_3\underset{\underset{\displaystyle CH_3}{|}}{C}ClCH_2Cl$$

$$CH_3\underset{\underset{\displaystyle CH_2Cl}{|}}{CH}CH_2Cl$$

19. (a) $CH_4 + Cl_2 \xrightarrow{\text{light}} CH_3Cl + HCl$

    (b) $CH_3Cl + Cl_2 \xrightarrow{\text{light}} CH_2Cl_2 + HCl$

          $CH_2Cl_2 + Cl_2 \xrightarrow{\text{light}} CHCl_3 + HCl$

    (c) $CH_3CH_2CH_3 + Br_2 \xrightarrow{\text{light}} CH_3CH_2CH_2Br + HBr$

         (an isomer, $CH_3CHBrCH_3$, will also be formed and in greater amount)

    (d) $CH_3CH_2CH_3 + Br_2 \xrightarrow{\text{light}} CH_3CHBrCH_3 + HBr$

         (a lesser amount of an isomer, $CH_3CH_2CH_2Br$, will also be formed)

20.

Structure	Common Name	IUPAC Name
(a) $CH=CH_2$	ethylene	ethene
(b) $CH \equiv CH$	acetylene	ethyne
(c) CH_3CH_2Cl	ethyl chloride	chloroethane

(d) CH_3CH_2OH ethyl alcohol   ethanol

(e) CH_3OCH_3 dimethyl ether   methoxymethane

(f) CH_3CHO acetaldehyde   ethanal

(g) CH_3COOH acetic acid   ethanoic acid

(h) $HCOOCH_3$ methyl formate   methylmethanoate

21. See Exercise 20 for common and IUPAC names.

22. Compound pairs (d) and (e), as well as (g) and (h), are isomers.

23. (a)   Chloromethane CH_3Cl

     (b)   Vinyl chloride $CH_2=CHCl$

     (c)   Chloroform $CHCl_3$

     (d)   Hexachloroethane CCl_3CCl_3

     (e)   Iodoethyne $CH{\equiv}CI$

     (f)   1-bromo-4-methyl- $BrCH_2CH=CHCHC{\equiv}CH$
           2-ene-5-hexyne |
                                                     CH_3

     (g)   1,1-dibromoethene $CH_2=CBr_2$

     (h)   1,2-dibromoethene $CHBr=CHBr$

24. Structure of vinyl acetylene, C_4H_4, $CH_2 = CH - C \equiv CH$

25. Formula of C_6H_8

    The formula would be C_6H_{14} if the compound were a saturated hydrocarbon.   This formula is 6 H atoms short of being saturated.   Removing two H atoms from a saturated hydrocarbon forms a carbon–carbon double bond.   Removing four H atoms can form a carbon–carbon triple bond. Therefore, C_6H_8 can contain three carbon–carbon double bonds or one double bond and one triple bond.

26. (a)   2,5-dimethyl-3-hexene   (d)   1,2-diphenylethene

         $CH_3\,CHCH=CHCHCH_3$
                 | |
                CH_3 CH_3

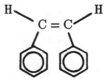

(b) 2–ethyl–3–methyl–1–pentene

$$CH_2= \overset{\overset{\displaystyle CH_2CH_3}{|}}{C}-\underset{\underset{\displaystyle CH_3}{|}}{C}HCH_2CH_3$$

(c) 4–methyl–2–pentene

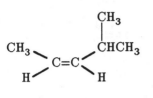

(e) 3–penten–1–yne

$$CH \equiv CCH = CHCH_3$$

(f) 3–phenyl–1–butyne

$$CH \equiv CCHCH_3$$

(g) vinyl bromide

$$CH_2 = CHBr$$

27. (a) $CH_3CH=\overset{\overset{}{|}}{C}CH_2CH_2CH_3$ 3–methyl–2–hexene
       $\underset{\displaystyle CH_3}{}$

    (b) $CH_3\overset{}{C} = \overset{}{C}CH_3$ 2,3–dimethyl–2–butene
         $\underset{\displaystyle H_3C \quad CH_3}{}$

    (c) $CH_3CH_2\,\overset{}{C}HCH=CH_2$ 3–isopropyl–1–pentene
         $\underset{\displaystyle CH(CH_3)_2}{}$

    (d) $CH_3CH_2CH=\overset{}{C}CH_2CH_3$ 3–methyl–3–hexene
         $\underset{\displaystyle CH_3}{}$

    (e) $CH_2C \equiv CH$ 3–phenyl–1–propyne

28. (a) All the pentynes

    1–pentyne   2–pentyne   3–methyl–1–butyne

    $CH_3CH_2CH_2C{\equiv}CH$ $CH_3CH_2C{\equiv}CCH_3$ $CH_3\,\overset{}{C}HC{\equiv}CH$
                                                          $\underset{\displaystyle CH_3}{}$

(b)   All the hexynes

1-hexyne

$CH_3CH_2CH_2CH_2C \equiv CH$

2-hexyne

$CH_3CH_2CH_2C \equiv CCH_3$

3-hexyne

$CH_3CH_2C \equiv CCH_2CH_3$

3-methyl-1-pentyne

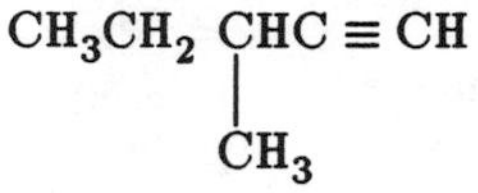

$$CH_3CH_2 \ \underset{\underset{CH_3}{|}}{C}HC \equiv CH$$

4-methyl-1-pentyne

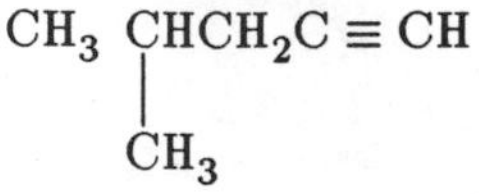

$$CH_3 \ \underset{\underset{CH_3}{|}}{C}HCH_2C \equiv CH$$

4-methyl-2-pentyne

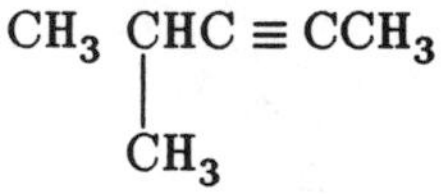

$$CH_3 \ \underset{\underset{CH_3}{|}}{C}HC \equiv CCH_3$$

3,3-dimethyl-1-butyne

$$CH_3 \ \underset{\underset{CH_3}{\overset{\overset{CH_3}{|}}{|}}}{C}C \equiv CH$$

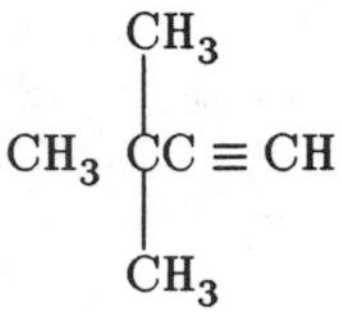

29. Complete the reactions and name the products.

(a) $CH_2=CHCH_3 \ + \ Br_2 \longrightarrow CH_2BrCHBrCH_3$ (1,2-dibromopropane)

(b) $CH_2=CH_2 \ + \ HBr \longrightarrow CH_3CH_2Br$ (bromoethane)

(c) $CH_3CH=CHCH_3 \ + \ H_2 \ \xrightarrow[\text{Pt, 25°C}]{\text{light}} \ CH_3CH_2CH_2CH_3$ (butane)

(d) $CH_2=CH_2 \ +H_2O \ \xrightarrow{H^+} \ CH_2CH_2OH$ (ethanol)

(e) $CH \equiv CH \ + \ 2 \ Br_2 \longrightarrow CHBr_2CHBr_2$ (1,1,2,2-tetrabromoethane)

30. $CaO \ + \ 3 \ C \ \xrightarrow[\text{furnace}]{\text{2500°C}} \ CaC_2 \ + \ CO$

$CaC_2 \ + \ 2 \ H_2O \longrightarrow CH \equiv CH \ + \ Ca(OH)_2$

$2 \ CH_4 \ \xrightarrow{\text{1500°C}} \ CH \equiv CH \ + \ 3 \ H_2$

31. Names of aromatic compounds

(a)

phenol

(g)

Cl

Cl

para–dichlorobenzene
(1,4–dichlorobenzene)

(b)

CH₃

toluene

(h)

OH

NO₂

para–nitrophenol

(c)

COOH

benzoic acid

(i)

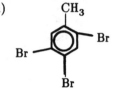

2,4,5–tribromotoluene

(d)

NH₂

aniline

(j)

CH₂CH₃

ethylbenzene

(e)

Cl

Cl

ortho–dichlorobenzene
(1,2–dichlorobenzene)

(k)

CH₃

OH

para–methylphenol

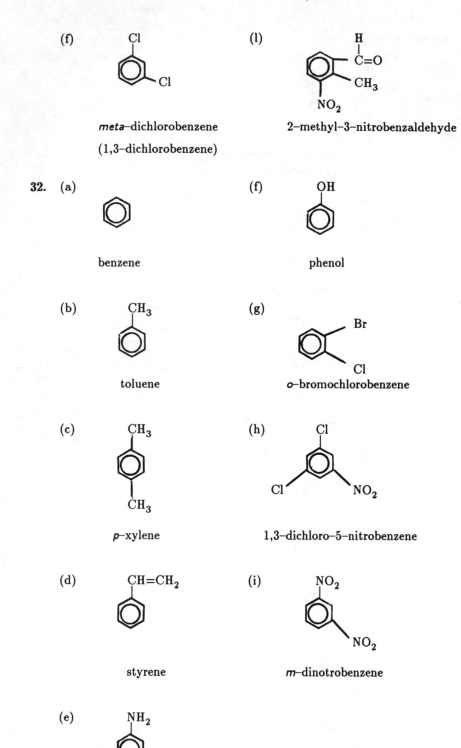

(f)

meta–dichlorobenzene

(1,3–dichlorobenzene)

(l)

2–methyl–3–nitrobenzaldehyde

32. (a)

benzene

(f)

phenol

(b)

toluene

(g)

o–bromochlorobenzene

(c)

p–xylene

(h)

1,3–dichloro–5–nitrobenzene

(d)

styrene

(i)

m–dinotrobenzene

(e)

aniline

33. (a) ethylbenzene

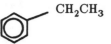

(b) benzoic acid

(c) 1,3,5–tribromobenzene

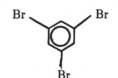

(d) naphthalene

(e) anthracene

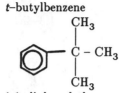

(f) *t*–butylbenzene

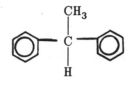

(g) 1,1–diphenylethane

34. (a) trichlorobenezene

1,2,3–trichlorobenzene

1,2,4–trichlorobenzene

1,3,5–trichlorobenzene

(b) dichlorobromobenzene

1,2–dichloro–3–bomobenzene

1,2–dichloro–4–bromobenzene

1,3–dichloro–2–bromo–
benzene

1,3–dichloro–4–bromobenzene

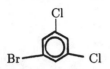

1,3–dichloro–5–bromo–
benzene

1,4–dichloro–2–bromobenzene

(c)   the benzene derivatives of formula C_8H_{10}

o–xylene or
1,2–dimethylbenzene

m–xylene or
1,3–methylbenzene

p–xylene or
1,4–dimethyl benzene

ethylbenzene

35. (a)   2–butanol   secondary

(b)   5–methyl–2–hexanol   secondary

(c)   3,7–dimethyl–4–nonanol   secondary

(d)   3–isopropyl–1–hexanol   primary

36. (8) Isomeric alcohols, formula $C_5H_{11}OH$

$CH_3CH_2CH_2CH_2CH_2OH$	1–pentanol	1°
$CH_3CH_2CH_2\underset{\underset{OH}{|}}{C}HCH_3$	2–pentanol	2°
$CH_3CH_2\underset{\underset{OH}{|}}{C}HCH_2CH_3$	3–pentanol	2°
$CH_3CH_2\underset{\underset{CH_3}{|}}{C}HCH_2OH$	2–methyl–1–butanol	1°
$CH_3\underset{\underset{CH_3}{|}}{C}HCH_2CH_2OH$	3–methyl–1–butanol	1°
$CH_3\overset{\overset{OH}{|}}{\underset{\underset{CH_3}{|}}{C}}CH_2CH_3$	2–methyl–2–butanol	3°
$CH_3\overset{\overset{OH}{|}}{C}H\underset{\underset{CH_3}{|}}{C}HCH_3$	3–methyl–2–butanol	2°
$CH_3\overset{\overset{CH_3}{|}}{\underset{\underset{CH_3}{|}}{C}}CH_2OH$	2,2–dimethyl–1–propanol	1°

37. Structural formulas

    (a)   2–pentanol $CH_3CH_2CH_2\underset{\underset{OH}{|}}{C}HCH_3$

    (b)   isopropyl alcohol $CH_3\underset{\underset{OH}{|}}{C}HCH_3$

(c)   2,2–dimethyl–1–heptanol

$$HOCH_2 \overset{\overset{\displaystyle CH_3}{|}}{\underset{\underset{\displaystyle CH_3}{|}}{C}} CH_2CH_2CH_2CH_2CH_3$$

(d)   1,3–propanediol $HOCH_2CH_2CH_2OH$

(e)   glycerol $HOCH_2 \overset{}{\underset{\underset{\displaystyle OH}{|}}{C}}HCH_2OH$

38. (a) $CH_3\overset{}{\underset{\underset{\displaystyle Br}{|}}{C}}HCH_3$ +   NaOH $\longrightarrow$ $CH_3\overset{}{\underset{\underset{\displaystyle OH}{|}}{C}}HCH_3$ +   NaBr

        isopropyl bromide   isopropyl alcohol

(b) $BrCH_2CH_2 \overset{}{\underset{\underset{\displaystyle CH_3}{|}}{C}}HCH_3$ + NaOH $\longrightarrow$ $HOCH_2CH_2\overset{}{\underset{\underset{\displaystyle CH_3}{|}}{C}}HCH_3$ + NaBr

        3–methyl–1–bromobutane

(c) $BrCH_2CH_2CH_2Br$ + NaOH $\longrightarrow$ $HOCH_2CH_2CH_2OH$ + 2 NaBr

        1,3–dibromopropane

39. (a) $CH_3CH_2CH_2OH$ $\xrightarrow[\Delta]{H^+,\ K_2Cr_2O_7}$ CH_3CH_2COOH + H_2O

                        (propanoic acid)

(b) $CH_3CH(OH)CH_3$ $\xrightarrow[\Delta]{H^+,\ K_2Cr_2O_7}$ $CH_3 \overset{\overset{\displaystyle }{}}{\underset{\underset{\displaystyle O}{\|}}{C}}CH_3$ + H_2O

                        (acetone)

(c) $CH_3\overset{\overset{\displaystyle CH_3}{|}}{\underset{\underset{\displaystyle OH}{|}}{C}}CH_2CH_3$ $\xrightarrow[\Delta]{H^+,\ K_2Cr_2O_7}$ no reaction (3° ROH)

(d)   2 $CH_3CH_2CH_2OH$ + 2 Na $\longrightarrow$ 2 $CH_3CH_2CH_2O^-Na^+$ + H_2

                        (sodium propoxide)

40. (a) $CH_3CH=CHCH_3 + H_2O \xrightarrow{H^+} CH_3CH_2CH(OH)CH_3$

(b) $CH_3CH=CHCH_3 + HBr \longrightarrow CH_3CH_2CHBrCH_3$

(c) $CH_2=CH_2 + H_2O \xrightarrow{H^+} CH_3CH_2OH$

$$CH_3CH_2OH \xrightarrow[\Delta]{H^+, K_2Cr_2O_7} \quad CH_3\overset{\displaystyle H}{\underset{|}{C}}=O$$

(d) $CH_3CH_2CH(OH)CH_3 \xrightarrow[\Delta]{H^+, K_2Cr_2O_7} CH_3CH_2\overset{\displaystyle O}{\overset{||}{C}}CH_3$

41. The molar mass of myricyl alcohol, an open chain saturated alcohol, contains 30 carbon atoms. The first three alcohols in the homologous series, CH_3OH, C_2H_5OH, and C_3H_7OH, leads to the formula $C_{30}H_{61}OH$.
molar mass = $(30)(12.01) + (62)(1.008) + (1)(16.00) = 438.8$

42. Ethylene glycol is superior to methyl alcohol as an antifreeze because of its low volatility. Methyl alcohol is much more volatile than water. If the radiator leaks gas under pressure (normally steam), it would primarily leak methanol vapor, thus eliminating the antifreeze. Ethylene glycol has a lower volatility than the water present, so it does not present this problem.

43. (a) Methanol, taken internally, is poisonous and capable of causing blindness and death. Breathing methanol vapors is also very dangerous.

(b) Physiologically, ethanol acts as a food, as a drug, and as a poison. It is a food, in a limited sense. The body is able to metabolize small amounts of it to carbon dioxide and water, resulting in the production of energy. As a drug, ethanol is often mistakenly considered to be a stimulant, but it is, in fact, a depressant. In moderate quantities, ethanol causes drowsiness and depresses brain fuctions. In larger quantities, ethanol causes nausea, vomiting, impaired perception, and a lack of coordination. Consumption of very large quantities may cause unconsciousness, and even death.

44. (a) methyl ethyl ether $\quad\quad CH_3CH_2OCH_3$

(b) dimethyl ether $\quad\quad CH_3OCH_3$

(c) methyl ethyl ether $\quad\quad CH_3CH_2OCH_3$

45. Ethers having the formula $C_5H_{12}O$ are: (IUPAC name in parentheses)

$CH_3OCH_2CH_2CH_2CH_3 \quad\quad$ methyl n–butyl ether
$\quad\quad\quad\quad\quad\quad\quad\quad\quad\quad\quad\quad$ (1–methoxybutane)

Chapter 20

$CH_3CH_2OCH_2CH_2CH_3$ ethyl *n*-propyl ether
(1-ethoxypropane)

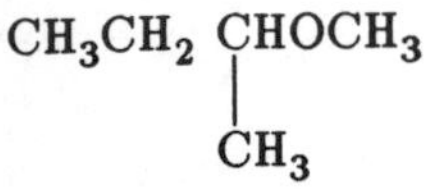

$CH_3CH_2\ CHOCH_3$
$|$
CH_3
    methyl *s*-butyl ether
(2-methoxybutane)

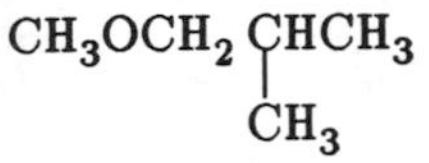

$CH_3OCH_2\ CHCH_3$
$|$
CH_3
    methyl isobutyl ether

(2-methyl-1-methoxyprpoane)

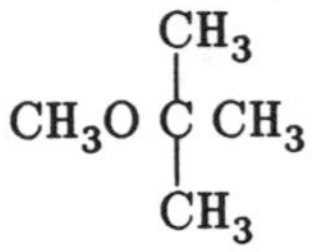

CH_3
$|$
$CH_3O\ C\ CH_3$
$|$
CH_3
    methyl *t*-butyl ether

(2-methyl-2-methoxypropane)

$CH_3CH_2OCH(CH_3)_2$ ethyl isopropyl ether
(2-ethoxypropane)

46. Williamson synthesis of ethers

(a) $CH_3CH_2OCH_3$ $CH_3CH_2ONa\ +\ CH_3Cl$

                       $CH_3CH_2Cl\ +\ CH_3ONa$

(b) $CH_3CH_2OCH_2CH_3$ $CH_3CH_2ONa\ +\ CH_3CH_2Cl$

(c) CH_3OCH_3 $CH_3ONa\ +\ CH_3Cl$

47. Propanal $CH_3CH_2C=O$
                      $|$
                     H
                     Propanone CH_3CCH_3
                                          $\overset{||}{O}$

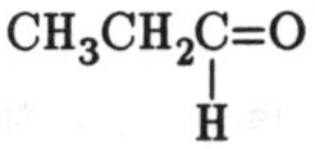

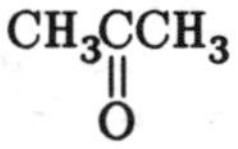

Propanal and propanone are isomers (C_3H_6O)

Butanal $CH_3CH_2CH_2C=O$ Butanone $CH_3CH_2\ CCH_3$
                          H O

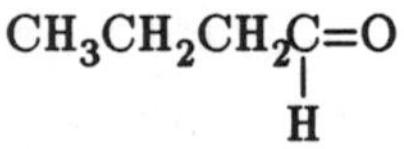

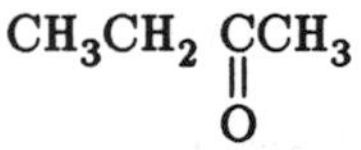

Butanal and butanone are isomers (C_4H_8O). Aldehydes and ketones of the same carbon content are isomeric compounds.

48. Names of aldehydes(common name in parentheses)

(a) $H_2C=O$ methanal (formaldehyde)

(b) $CH_3CH_2CH_2\overset{|}{\underset{H}{C}}=O$ butanal (butyraldehyde)

(c) $CH_3\overset{|}{\underset{CH_3}{C}}HCH_2\overset{|}{\underset{H}{C}}=O$ 3–methylbutanal

(d)   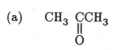$\overset{H}{\underset{}{|}}$
                                     $C=O$ benzaldehyde

(e) $O=\overset{|}{\underset{H}{C}}CH_2CH_2\overset{||}{\underset{O}{C}}-H$ butanedial

(f) $\overset{Cl}{\underset{}{}}$ $\overset{O}{\overset{||}{C}}-H$

                                                                        2–chloro–5–isopropyl
                                                                        benzaldehyde

        CH_3 $\overset{CH}{}$ CH_3

(g) $CH_3\overset{|}{\underset{OH}{C}}HCH_2\overset{O}{\overset{||}{C}}-H$ 3–hydroxybutanal

49. Names of ketones

(a) $CH_3\overset{}{\underset{O}{\overset{||}{C}}}CH_3$ propanone, acetone, dimethyl ketone

(b) $CH_3CH_2\overset{O}{\overset{||}{C}}CH_3$ 2–butanone, methyl ethyl ketone

(c) $\overset{O}{\overset{||}{C}}CH_2CH_3$ 1–phenyl–1–propanone, phenyl
                                                                        ethyl ketone, propiophenone

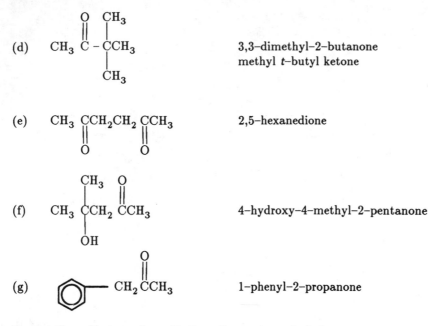

(d) $CH_3 \overset{O}{\underset{}{C}} - \overset{CH_3}{\underset{CH_3}{C}}CH_3$ 3,3–dimethyl–2–butanone
          methyl *t*–butyl ketone

(e) $CH_3 \overset{}{\underset{O}{C}}CH_2CH_2 \overset{}{\underset{O}{C}}CH_3$ 2,5–hexanedione

(f) $CH_3 \overset{CH_3}{\underset{OH}{C}}CH_2 \overset{O}{\underset{}{C}}CH_3$ 4–hydroxy–4–methyl–2–pentanone

(g) $CH_2\overset{O}{\underset{}{C}}CH_3$ 1–phenyl–2–propanone

50.   Preparation of ketones by oxidation of secondary alcohols.

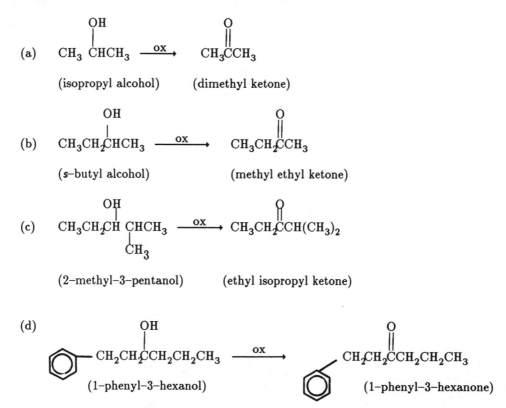

(a) $CH_3 \overset{OH}{\underset{}{C}}HCH_3 \xrightarrow{\text{ox}} CH_3\overset{O}{\underset{}{C}}CH_3$

          (isopropyl alcohol)   (dimethyl ketone)

(b) $CH_3CH_2\overset{OH}{\underset{}{C}}HCH_3 \xrightarrow{\text{ox}} CH_3CH_2\overset{O}{\underset{}{C}}CH_3$

          (s–butyl alcohol)   (methyl ethyl ketone)

(c) $CH_3CH_2\overset{OH}{\underset{CH_3}{C}}H \overset{}{C}HCH_3 \xrightarrow{\text{ox}} CH_3CH_2\overset{O}{\underset{}{C}}CH(CH_3)_2$

          (2–methyl–3–pentanol)   (ethyl isopropyl ketone)

(d) $CH_2CH_2\overset{OH}{\underset{}{C}}CH_2CH_2CH_3 \xrightarrow{\text{ox}} CH_2CH_2\overset{O}{\underset{}{C}}CH_2CH_2CH_3$

          (1–phenyl–3–hexanol)   (1–phenyl–3–hexanone)

51. (a) Aldehydes produce a positive Tollens' test.

(b) The formation of a silver mirror is the visible evidence for the Tollens test.

(c)
$$CH_3 \overset{\overset{\displaystyle H}{|}}{C}{=}O \ + \ 2\ Ag^+ + NH_3 \longrightarrow CH_3COONH_4 \ + \ 2\ Ag \ (mirror)$$

52. Chemical tests to distinguish between pairs of compounds.

(a) $CH_3CH_2OCH_2CH_3 \ + \ Na \longrightarrow$ no reaction

$2\ CH_3OH \ + \ 2\ Na \longrightarrow 2\ CH_3ONa \ + \ H_2$ (evolution of gas)

(b) $CH_3CHO \ + \ 2Ag^+ \xrightarrow[H_2O]{NH_3} CH_3COO^-NH_4^+ \ + \ 2\ Ag$
(silver mirror)

$$CH_3\overset{\overset{\displaystyle O}{\|}}{C}CH_3 \ + \ 2\ Ag^+ \xrightarrow[H_2O]{NH_3} \text{no reaction}$$

(c) $CH_3(CH_2)_5CH{=}CH_2 \ + \ Br_2 \longrightarrow CH_3(CH_2)_5\overset{\overset{\displaystyle Br}{|}}{C}HCH_2Br$

(rapid decolorizing of bromine)

$CH_3(CH_2)_6CH_3 \ + \ Br_2 \longrightarrow$ no reaction (light is required)

(d) $CH_3(CH_2)_3CH_2OH$ and CH_3CH_2COOH

Test with blue litmus paper. The acid will turn the paper red. The alcohol will not change the color of the blue litmus.

(e) $CH_2{=}CHOCH{=}CH_2$ and $CH_3CH_2OCH_2CH_3$

Divinyl ether decolorizes bromine and diethyl ether does not.

53. Carboxylic acids, common name, IUPAC name

$HCOOH$	formic acid; methanoic acid
CH_3COOH	acetic acid; ethanoic acid
CH_3CH_2COOH	propionic acid; propanoic acid
$CH_3CH_2CH_2COOH$	butyric acid; butanoic acid
$CH_3CH_2CH_2CH_2COOH$	valeric acid; pentanoic acid

54. Lauric acid $\qquad$ $CH_3(CH_2)_{10}COOH$

Myristic acid $\qquad$ $CH_3(CH_2)_{12}COOH$

Palmitic acid $\qquad$ $CH_3(CH_2)_{14}COOH$

Stearic acid $\qquad$ $CH_3(CH_2)_{16}COOH$

55. $CH_3(CH_2)_4COOH$ $\qquad$ hexanoic acid

$CH_3CH_2CH_2\underset{\underset{CH_3}{|}}{C}HCOOH$ $\qquad$ 2–methylpentanoic acid

$\underset{\underset{CH_3}{|}}{C}H_3CHCH_2CH_2COOH$ $\qquad$ 4–methylpentanoic acid

$CH_3CH_2\underset{\underset{CH_3}{|}}{C}HCH_2COOH$ $\qquad$ 3–methypentanoic acid

$CH_3CH_2\overset{\overset{CH_3}{|}}{\underset{\underset{CH_3}{|}}{C}}COOH$ $\qquad$ 2,2–dimethylbutanoic acid

$CH_3\overset{\overset{CH_3}{|}}{\underset{\underset{CH_3}{|}}{C}}CH_2COOH$ $\qquad$ 3,3–dimethylbutanoic acid

$CH_3\overset{\overset{CH_3}{|}}{C}H\underset{\underset{CH_3}{|}}{C}HCOOH$ $\qquad$ 2,3–dimethylbutanoic acid

$CH_3CH_2\underset{\underset{CH_2CH_3}{|}}{C}HCOOH$ $\qquad$ 2–ethylbutanoic acid

56. (a) $\quad CH_3CHBrCOOH$ $\qquad$ 2–bromopropanoic acid

(b) $\quad CH_2=CHCH_2COOH$ $\qquad$ 3–butenoic acid

(c) $\quad CH_3\underset{\underset{CH_2CH_3}{|}}{C}HCOOH$ $\qquad$ 2–methylbutanoic acid

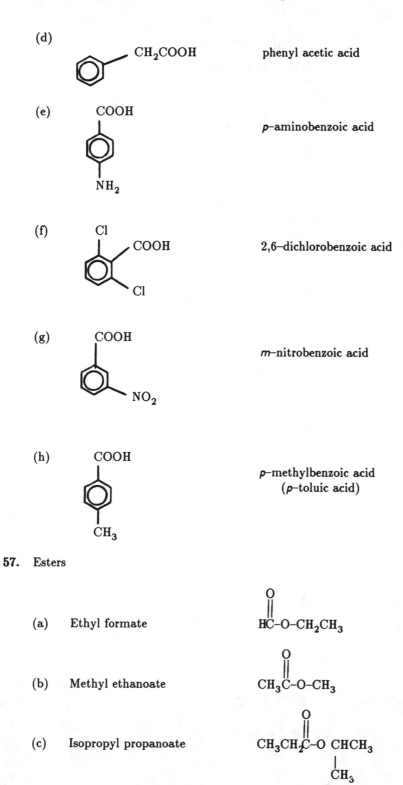

(d) CH$_2$COOH   phenyl acetic acid

(e) COOH   p–aminobenzoic acid
NH$_2$

(f) Cl COOH   2,6–dichlorobenzoic acid
Cl

(g) COOH   m–nitrobenzoic acid
NO$_2$

(h) COOH   p–methylbenzoic acid
(p–toluic acid)
CH$_3$

57. Esters

(a) Ethyl formate
$$\underset{\text{O}}{\overset{\text{O}}{\underset{||}{}}}$$
HC–O–CH$_2$CH$_3$

(b) Methyl ethanoate
CH$_3$C–O–CH$_3$

(c) Isopropyl propanoate
CH$_3$CH$_2$C–O CHCH$_3$
CH$_3$

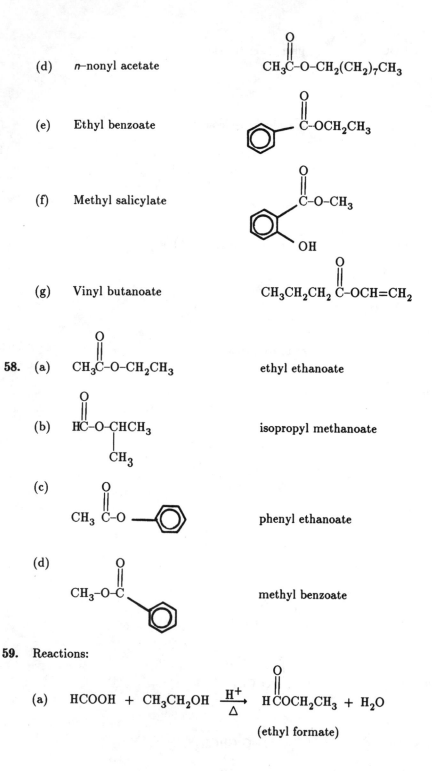

(d)   *n*–nonyl acetate

(e)   Ethyl benzoate

(f)   Methyl salicylate

(g)   Vinyl butanoate

58.   (a)   ethyl ethanoate

(b)   isopropyl methanoate

(c)   phenyl ethanoate

(d)   methyl benzoate

59.   Reactions:

(a) $HCOOH + CH_3CH_2OH \xrightarrow[\Delta]{H^+} H\overset{O}{\underset{}{\overset{\|}{C}}}OCH_2CH_3 + H_2O$

(ethyl formate)

(b) $CH_3CH_2COOH + CH_3OH \xrightarrow[\Delta]{H^+}$ $CH_3CH_2\overset{\displaystyle O}{\overset{\|}{C}}OCH_3 + H_2O$

(methyl propanoate)

(c) 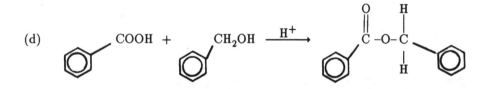 COOH $+ CH_3CH_2CH_2OH \xrightarrow[\Delta]{H^+}$ $\overset{\displaystyle O}{\overset{\|}{C}}OCH_2CH_2CH_3 + H_2O$

(n–propyl benzoate)

60. Reactions:

(a) $CH_3COOH + NaOH \longrightarrow CH_3COO^-Na^+ + H_2O$

(b) $CH_3\overset{\displaystyle O}{\overset{\|}{C}}H\overset{\displaystyle }{C}OH + NH_3 \longrightarrow$ $CH_3CHCOO^-NH_4^+$

  $\overset{\displaystyle }{\underset{OH}{|}}$ $\overset{\displaystyle }{\underset{OH}{|}}$

(c) $2\ CH_3COOH + 2\ K \longrightarrow 2\ CH_3COOK + H_2$

(d) COOH $+$ $CH_2OH \xrightarrow{H^+}$ $\overset{\displaystyle O}{\overset{\|}{C}} - O - \overset{\displaystyle H}{\underset{H}{\overset{|}{C}}}$

61. The correct statements are b, f, h, j, k, l, n, o, p, q, r, t, v, x, z.

(a) In the alkane homologous series, the formula of each member differs from the preceding member by CH_2.

(c) The IUPAC name for $CH_3CH_2CH_2CHClCH_3$ is 2–chloropentane.

(d) Propane is not an isomer of propene.

(e) Isopropyl alcohol is an isomer of methyl ethyl ether.

(g) Isobutane and methylpropane are both correct names for the same compound.

(i) Two monosubstituted products will result from the chlorination of butane.

(m) Dimethyl ketone, acetone, and propanone all have the same molecular and structural formulas.

(s) Ethanal may not be distinguished from propanal by use of Tollens' reagent.

(u) The chemical characteristics of CH_3OH and $NaOH$ are different.

(w) Aldehydes are easily oxidized.

(y) Formic acid is structurally the simplest carboxylic acid.

62. (a) $+ CH_2 - CH_2 +_n$ polyethylene

(b) $+ CH_2 - \underset{\underset{\displaystyle Cl}{|}}{CH} +_n$ polyvinyl chloride

(c) $+ CH_2 - \underset{\underset{\displaystyle C_6H_5}{|}}{CH} +_n$ polystyrene

(d) $+ CH_2 - CCl_2 +_n$ saran

(e) $+ CH_2 - \underset{\underset{\displaystyle CN}{|}}{CH} +_n$ polyacrylonitrile

(f) $+ CF_2 - CF_2 +_n$ teflon

63. (a) $+ CH_2 - \underset{\underset{\displaystyle CH_3}{|}}{CH} - CH_2 - \underset{\underset{\displaystyle CH_3}{|}}{CH} - CH_2 - \underset{\underset{\displaystyle CH_3}{|}}{CH} - CH_2 - \underset{\underset{\displaystyle CH_3}{|}}{CH} +_n$

$+ CH_2 - \underset{\underset{\underset{\displaystyle CH_3}{|}}{\underset{\displaystyle CH_2}{|}}}{CH} - CH_2 - \underset{\underset{\underset{\displaystyle CH_3}{|}}{\underset{\displaystyle CH_2}{|}}}{CH} - CH_2 - \underset{\underset{\underset{\displaystyle CH_3}{|}}{\underset{\displaystyle CH_2}{|}}}{CH} - CH_2 - \underset{\underset{\underset{\displaystyle CH_3}{|}}{\underset{\displaystyle CH_2}{|}}}{CH} +_n$

$+ CH - \underset{\overset{\overset{\displaystyle CH_3}{|}}{}}{CH} - \underset{\overset{\overset{\displaystyle CH_3}{|}}{}}{CH} - CH - \underset{\overset{\overset{\displaystyle CH_3}{|}}{}}{CH} - \underset{\overset{\overset{\displaystyle CH_3}{|}}{}}{CH} - CH - \underset{\underset{\displaystyle CH_3}{|}}{CH} +_n$

(b)   Monomer molar mass of ethylene = 28.05

$$\frac{\text{molar mass of polymer}}{\text{molar mass of monomer}} = \text{number of monomer units}$$

$$\frac{35000}{28.05} = 1250 \text{ units}$$

INTRODUCTION TO BIOCHEMISTRY

1. In Table 21.1, the sweetest disaccharide is sucrose. The sweetest monosaccharide is fructose.

2. Fatty acids in vegetable oils are more unsaturated than fatty acids in animal fats. This is because vegetable oils contain higher percentages of oleic and linoleic (unsaturated) acids than animal fats.

3. Of the common amino acids listed in Table 21.3, two, aspartic acid and glutamic acid, have more than one carboxyl group. The following amino acids have more than one amino group: arginine, histidine, lysine, and tryptophan.

4. There are three disulfide linkages in each molecule of beef insulin.

5. In DNA, the nitrogen bases are off to the side while the deoxyribose and phosphoric acid are part of the backbone chain.

6. In the double stranded helix structure of DNA (Figure 21.6), the hydrogen bonding of the nitrogen bases is as follows: guanine and cytosine are mutually bonded to each other as are adenine and thymine.

7. The four major classes of biomolecules are carbohydrates, lipids, proteins, and nucleic acids.

8. Monosaccharides, disaccharides, and polysaccharides. The simplest type of carbohydrate is the monosaccharide.

9. An aldose is a monosaccharide containing an aldehyde group on one carbon atom and a hydroxyl group on each of the other carbon atoms. An aldotetrose is an aldose containing four carbon atoms. A ketose is a monosaccharide containing a ketone group on one carbon atom and a hydroxyl group on each of the other carbon atoms. A ketohexose is a ketone containing six carbon atoms. Examples: aldose, glucose; aldotetrose, erythrose; ketose, fructose; ketohexose, fructose.

10. Classification of saccharides.

Monosaccharides	Disaccharides	Polysaccharides
Glucose	Sucrose	Cellulose
Fructose	Maltose	Glycogen
Galactose	Lactose	Starch
Ribose		

11.
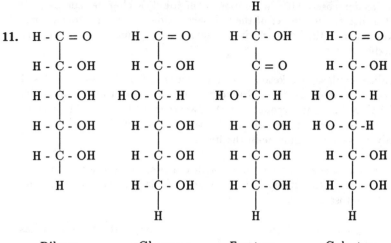

| Ribose | Glucose | Fructose | Galactose |

12. Cyclic structural formulas

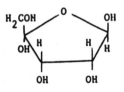

Ribose

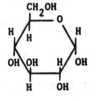

Glucose

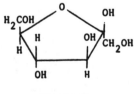

Fructose

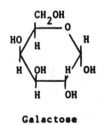

Galactose

13. Properties and sources

Ribose: Ribose is a white, crystalline, water soluble pentose sugar, present in adenosine triphosphate (ATP), one of the chemical energy carriers in the body. Ribose and one of its derivatives, deoxyribose, are also important components of the nucleic acids, DNA and RNA, the genetic information carriers in the body.

Glucose: Glucose is an aldohexose and is found in the free state in plant and animal tissues. Glucose is commonly known as dextrose . It is a component of the disaccharides sucrose, maltose, and lactose, and it is also the monomer of the polysaccharides starch, glycogen, and cellulose. Glucose is the key sugar of the body and is circulated by the blood stream to provide energy to all parts of the body.

Fructose: Fructose, also called levulose, is a ketohexose and occurs in fruit juices as well as in honey. Fructose is also a constituent of sucrose. Fructose is the sweetest of all sugars, having about twice the sweetness of glucose. This accounts for the sweetness of honey. The enzyme invertase, present in bees, splits sucrose into glucose and fructose. Fructose is metabolized directly, but it is also readily converted to glucose in the liver.

Galactose: Galactose is an aldohexose and occurs, along with glucose, in lactose and in many oligo- and polysaccharides, such as pectin, gums, and mucilages. Galactose is synthesized in the mammary glands to make the lactose of milk.

14. Lactic acid has a hydroxyl group and an acid group. It is not a carbohydrate, because it has neither an aldehyde nor a ketone group, and will not yield one upon hydrolysis.

15. The monosaccharide composition of:

    (a) Sucrose; a disaccharide made from one unit of glucose and one unit of fructose.

    (b) Maltose; a disaccharide made from two units of glucose.

    (c) Lactose; a disaccharide made from one unit of galactose and one unit of glucose.

    (d) Starch; a polysaccharide made from many units of glucose.

16. Cyclic structural formulas for sucrose and maltose.

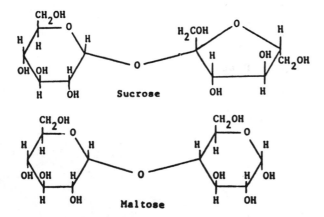

17. The formation of a disaccharide, $C_{12}H_{22}O_{11}$, from a monosaccharide, $C_6H_{12}O_6$, involves combining two monosaccharide units with a molecule of water splitting out between them.

18. (a) Maltose $\xrightarrow[H_2O]{\text{maltase}}$ Glucose + Glucose

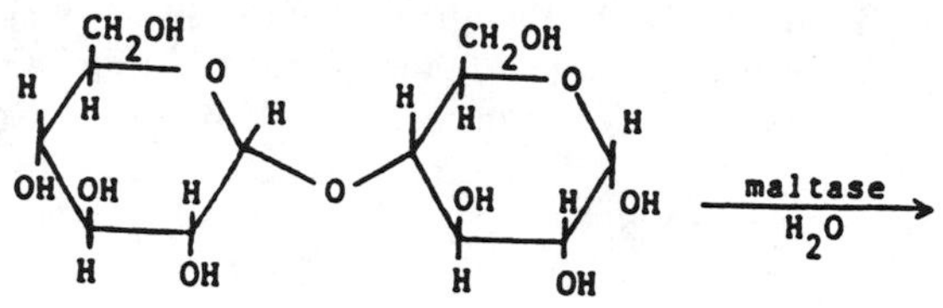

Maltose

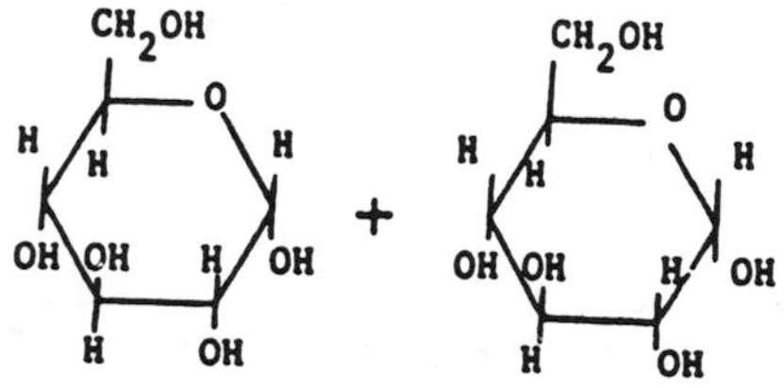

Glucose Glucose

(b) Sucrose $\xrightarrow[H_2O]{\text{sucrase}}$ Glucose + Fructose

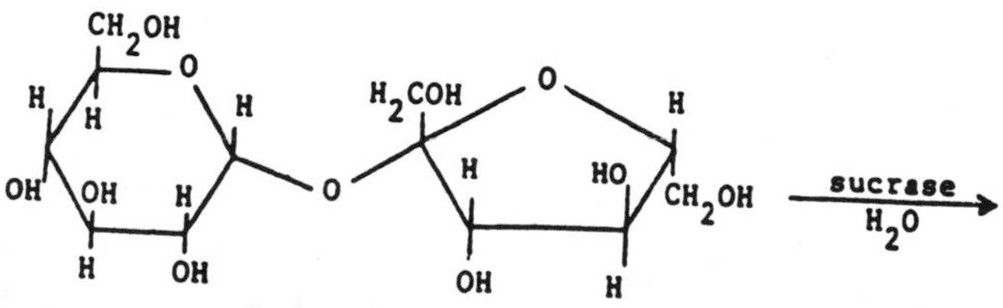

Sucrose

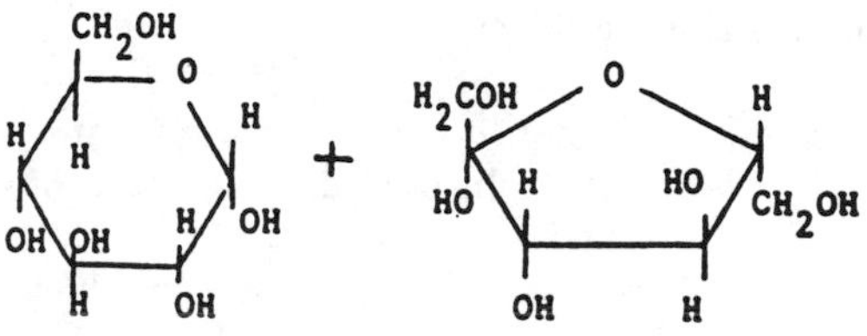

Glucose Fructose

(c) Sucrase catalyzes the hyrolysis of sucrose while maltase catalyzes the hydrolysis of maltose.

19. Starch and cellulose have their basic composition in common. They contain many glucose units joined in long chains forming polysaccharide molecules with a high molar mass. The difference in properties, which are highly significant, are due to the different manner in which the glucose units are attached to each other. The primary use of starch is for food. Man cannot utilize cellulose as food due to a lack of the necessary enzymes to hydrolyze cellulose to usable glucose.

20. Carbohydrates are stored in the human body as glycogen, a polysaccharide of glucose.

21. Carbohydrate metabolism

$$\text{Starch} \xrightarrow[\text{mouth}]{\text{amylase}} \text{Dextrins} \xrightarrow[\text{intestines}]{\text{amylase}} \text{Maltose}$$

$$\text{Maltose} \xrightarrow[\text{intestines}]{\text{maltase}} \text{Glucose}$$

Glucose is absorbed through the intestinal walls into the blood stream. From there the glucose may be stored in the liver as glycogen or utilized as energy by oxidation to carbon dioxide and water.

Sugars, such as sucrose and lactose, are converted to monosaccharides by specific enzymes in the intestines.

22. Natural sources of sucrose, maltose, lactose, and starch are:

Sucrose: Sucrose is found in the free state throughout the plant kingdom. Sugar cane contains 15% to 20% sucrose, while sugar beets contain 10% to 17% sucrose. Maple syrup and sorghum are also good sources of sucrose.

Maltose: Maltose is found in sprouting grain, but occurs much less commonly in nature than either sucrose or lactose.

Lactose: Lactose, also known as milk sugar, is found free in nature mainly in the milk of mammals. Human milk contains about 6.7% lactose and cow milk about 4.5% of this sugar.

Starch: Starch is a polymer of glucose. It is found mainly in the seeds, roots, and tubers of plants. Corn, wheat, potatoes, rice, and cassava are the chief sources of starch.

23. Invert sugar is sweeter than sucrose, from which it comes, because fructose is far sweeter than sucrose. The relative sweetnesses are: fructose, 100; glucose, 43; sucrose, 58.

24. Substances are classified as lipids on the basis of their solubility in nonpolar solvents such as diethyl ether, benzene, and chloroform, their insolubility in water, and their greasy feeling.

25. Structural formulas

Glycerol

$$\begin{array}{l} CH_2OH \\ | \\ CHOH \\ | \\ CH_2OH \end{array}$$

Palmitic acid	$CH_3(CH_2)_{14}COOH$
Oleic acid	$CH_3(CH_2)_7CH=CH(CH_2)_7COOH$
Stearic acid	$CH_3(CH_2)_{16}COOH$
Linoleic acid	$CH_3(CH_2)_4CH=CHCH_2CH=CH(CH_2)_7COOH$

26. Fats are solid and vegetable oils are liquid at room temperature. Fats contain higher amounts of saturated fatty acids and vegetable oils contain higher amounts of unsaturated fatty acids.

27. A triacylglycerol (triglyceride) is a triester of glycerol. Most animal fats are triacylgycerols. Tristearin, in the next problem, is a triacylglycerol.

28. Tristearin

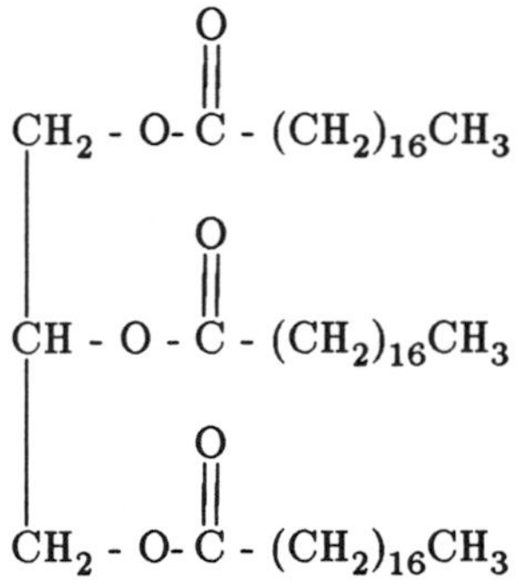

$$CH_2 - O - \overset{\overset{\displaystyle O}{\|}}{C} - (CH_2)_{16}CH_3$$

$$CH - O - \overset{\overset{\displaystyle O}{\|}}{C} - (CH_2)_{16}CH_3$$

$$CH_2 - O - \overset{\overset{\displaystyle O}{\|}}{C} - (CH_2)_{16}CH_3$$

29. Triacylglycerol

$$CH_2 - O - \overset{\overset{\displaystyle O}{\|}}{C} - (CH_2)_7CH=CHCH_2CH=CH(CH_2)_4CH_3 \text{ (linoleic)}$$

$$CH - O - \overset{\overset{\displaystyle O}{\|}}{C} - (CH_2)_{16}CH_3 \text{ (stearic)}$$

$$CH_2 - O - \overset{\overset{\displaystyle O}{\|}}{C} - (CH_2)_7CH=CH(CH_2)_7CH_3 \text{ (oleic)}$$

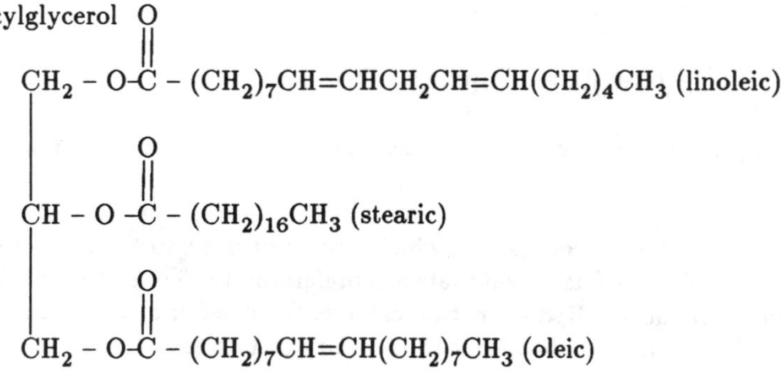

There would be two other formulas possible using these same three acids. In one, the linoleic acid would be in the middle and in the other, oleic acid would be in the middle. The top and bottom positions are equivalent.

30. (a) Tripalmitin is a fat in which all the fatty acid units are palmitic acid.

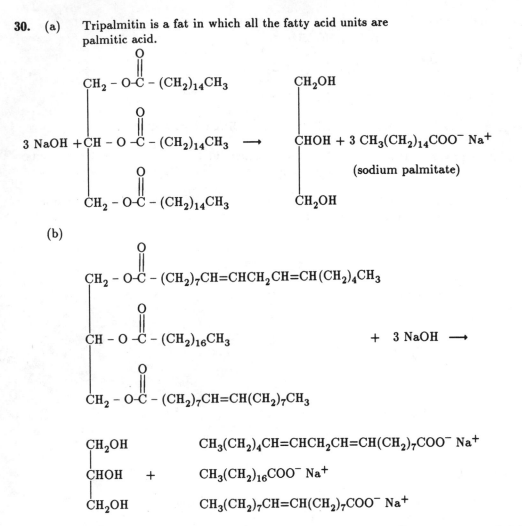

$$3 \ NaOH + \begin{array}{l} CH_2 - O - \overset{\overset{\displaystyle O}{\|}}{C} - (CH_2)_{14}CH_3 \\[2mm] CH - O - \overset{\overset{\displaystyle O}{\|}}{C} - (CH_2)_{14}CH_3 \\[2mm] CH_2 - O - \overset{\overset{\displaystyle O}{\|}}{C} - (CH_2)_{14}CH_3 \end{array} \longrightarrow \begin{array}{l} CH_2OH \\[2mm] CHOH + 3 \ CH_3(CH_2)_{14}COO^- \ Na^+ \\[1mm] \text{(sodium palmitate)} \\[2mm] CH_2OH \end{array}$$

(b)

$$\begin{array}{l} CH_2 - O - \overset{\overset{\displaystyle O}{\|}}{C} - (CH_2)_7 CH=CHCH_2 CH=CH(CH_2)_4 CH_3 \\[3mm] CH - O - \overset{\overset{\displaystyle O}{\|}}{C} - (CH_2)_{16}CH_3 \\[3mm] CH_2 - O - \overset{\overset{\displaystyle O}{\|}}{C} - (CH_2)_7 CH=CH(CH_2)_7 CH_3 \end{array} \qquad + \quad 3 \ NaOH \longrightarrow$$

$$\begin{array}{l} CH_2OH \\ CHOH \\ CH_2OH \end{array} \quad + \quad \begin{array}{l} CH_3(CH_2)_4 CH=CHCH_2 CH=CH(CH_2)_7 COO^- \ Na^+ \\ CH_3(CH_2)_{16}COO^- \ Na^+ \\ CH_3(CH_2)_7 CH=CH(CH_2)_7 COO^- \ Na^+ \end{array}$$

The fatty acids top to bottom are: sodium linoleate, sodium stearate, and sodium oleate.

31. Vegetable oils can be solidified by hydrogenation, which adds hydrogen to the double bonds, to saturate the double bonds and form fats. Solid fats are preferable to oils for the manufacture of soaps and for certain food products. Hydrogenation extends the shelf–life of oils, because it is oxidation at the points of unsaturation that leads to rancidity of fats and oils.

32. Fats are an important food source for man. They normally account for 25 to 50 percent of caloric intake. Fats are the major constituent of adipose tissue, which is distributed throughout the body. In addition to being a source of reserve energy, fat deposits function to insulate the body against loss of heat and protect vital organs against mechanical injury.

33. The three essential fatty acids are linoleic, linolenic, and arachidonic acids. Diets lacking these fatty acids lead to impaired growth and reproduction and skin disorders, such as eczema and dermatitis.

34. The structural formula of cholesterol is

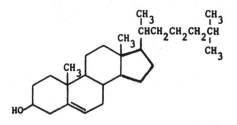

35. The ring structure common to all steroids is

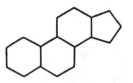

36. Some common foods with high (over 10%) protein content are gelatin, fish, beans, nuts, cheese, eggs, poultry, and meat of all kinds.

37. All amino acids contain a carboxylic acid group and an amino group.

38. The amino acids in proteins are called α–amino acids because the amine group is always attached in the α position, that is the first carbon next to the carboxylic acid group.

39. Dipeptides of glycine and phenylalanine

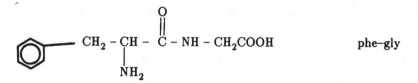

phe–gly

gly–phe

40. (a) Glycylglycine

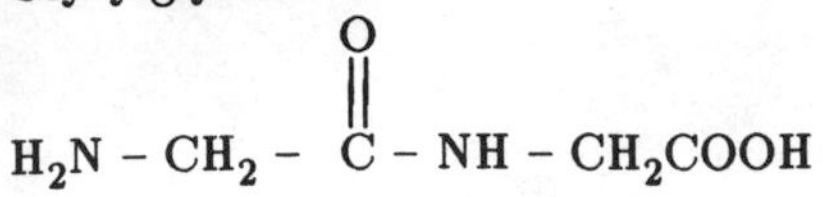

$$H_2N - CH_2 - \overset{\overset{\textstyle O}{\|}}{C} - NH - CH_2COOH$$

(b) Glycylglycylalanine

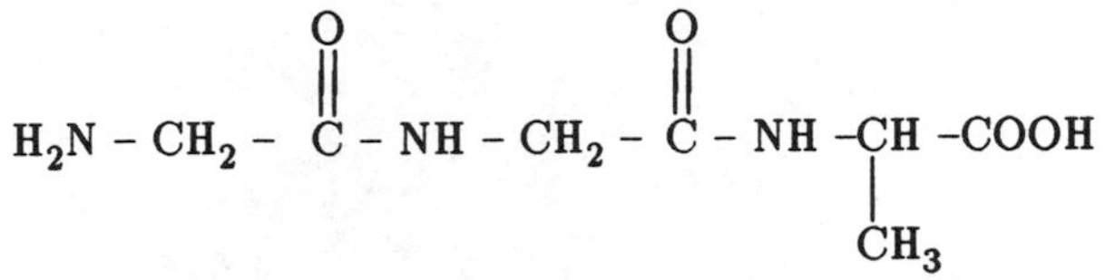

$$H_2N - CH_2 - \overset{\overset{\textstyle O}{\|}}{C} - NH - CH_2 - \overset{\overset{\textstyle O}{\|}}{C} - NH -\underset{\underset{\textstyle CH_3}{|}}{CH} -COOH$$

(c) Leucylmethionylglycylserine

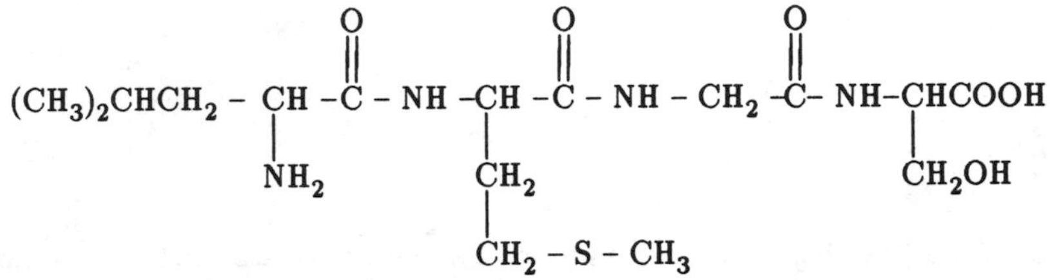

$$(CH_3)_2CHCH_2 - \underset{\underset{\textstyle NH_2}{|}}{CH} -\overset{\overset{\textstyle O}{\|}}{C} - NH -\underset{\underset{\textstyle CH_2}{|}}{CH} -\overset{\overset{\textstyle O}{\|}}{C} - NH - CH_2 -\overset{\overset{\textstyle O}{\|}}{C}- NH-\underset{\underset{\textstyle CH_2OH}{|}}{CH}COOH$$

$$CH_2 - S - CH_3$$

41. All possible tripeptides of glycine (gly), phenylalanine (phe), and leucine (leu).

gly – phe – leu	leu – phe – gly	gly – leu – phe
phe – gly – leu	leu – gly – phe	phe – leu –gly

42. Essential amino acids are those which are needed by the human body but cannot be synthesized by the body. Therefore, it is essential that they be included in the diet. They are: isoleucine, leucine, lysine, methionine, phenylalanine, threonine, tryptophan, and valine.

43. The proteins consumed by a human are converted by digestive enzymes into smaller peptides and amino acids. These smaller units are utilized in many ways:

(a) to replace and repair body tissues

(b) to synthesize new proteins.

(c) to synthesize other nitrogen–containing substances, such as enzymes, certain hormones, and heme molecules.

(d) to synthesize nucleic acids

(e) to synthesize other necessary foods, such as carbohydrates and fats.

Proteins are catabolized (degraded) to carbon dioxide, water, and urea. Urea, containing nitrogen, is eliminated from the body in the urine.

44. Tissue proteins are continuously being broken down and resynthesized. Protein is continually needed in a balanced diet because the body does not store free amino acids. They are needed to:

   (a) replace and repair body tissue

   (b) synthesize new proteins

   (c) synthesize other nitrogen–containing substances, such as enzymes, some hormones, and bone

   (d) synthesize nucleic acids

   (e) synthesize other necessary foods, such as carbohydrates and fats.

45. The component parts that make up DNA

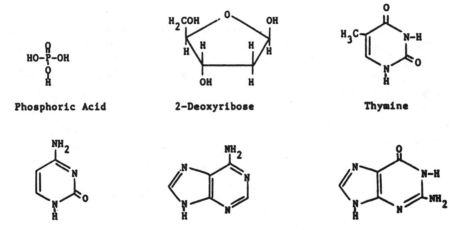

46. (a) The three units that make up a nucleotide are a phosphate, ribose or deoxyribose, and one of the four nitrogen–containing bases. (A, T, G, C)

   (b) In DNA, the four types of nucleotides are

   phosphate – deoxyribose – thymine
   phosphate – deoxyribose – cytosine
   phosphate – deoxyribose – adenine
   phosphate – deoxyribose – guanine

(c)   Structure and name of one of the nucleotides

phosphate – deoxyribose – cytosine

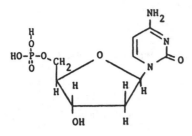

cytosine deoxyribonucleotide

47.   The structure of DNA, as proposed by Watson and Crick, is in the form of a double helix structure with both strands coiled around the same axis.   Along each stand, phosphate and deoxyribose units alternate.   Each deoxyribose unit has one of the bases attached, which is in turn hydrogen–bonded to a complementary base on the other strand.   Thus, the two strands are linked at each deoxyribose unit by two bases.

48.   The two helices of the double helix are joined together by hydrogen bonds between bases.   The structure of the bases is such that this hydrogen bonding is only possible to one specific base for each other base.   That is, adenine is always hydrogen–bonded to cytosine.   Therefore, the hydrogen bonding requires a certain structure on the adjoining helix.

49.   Complementary bases are the pairs that "fit" to each other by hydrogen bonds between the two helices of DNA.   For DNA, the complemetary pairs are thymine with adenine and cytocine with guanine, or TA and CG.

50.   If a segment of a DNA strand has base sequence C–G–A–T–T–G–C–A, the other strand of the double helix will have the base sequence G–C–T–A–A–C–G–T.

51.   Replication of DNA begins with the unwinding of the double helix at the hydrogen bonds between the bases to form two separate strands.   Each strand then combines with the proper free nucleotides to produce two identical replicas of the original double helix.   This replication of DNA occurs just before the cell divides, giving each daughter cell the full genetic code of the parent cell.

52.   DNA contains the genetic code of life.   For any individual, the sequence of bases and the length of the nucleotide chains in the DNA molecules contain the coded messages that determine all the characteristics of the individual, including the reproduction of that species.   Because of the mechanics of human reproduction, the offspring is a combination of the chromosomes of each parent, thus will not be a carbon copy of either parent.

53.   The differences between DNA and RNA are:

(a)   RNA exists in the form of a single–stranded helix, whereas DNA is a double helix.

(b)   RNA contains the pentose sugar ribose, whereas DNA contains deoxyribose.

(c)   RNA contains the base uracil, whereas DNA contains thymine.

54. In ordinary cell division, known as mitosis, each DNA molecule forms a duplicate by uncoiling to single strands.   Each strand then assembles the complementary portion from available free nucleotides to form duplicates of the original DNA molecule.

In most higher forms of life, reproduction takes place by the union of the male sperm with the female egg.   Cell splitting to form the sperm cell and the egg cell occurs by a different and more complicated process called meiosis.   In meiosis, the sperm cell carries only one half the chromosomes from its original cell, and the egg cell also carries one half of its original chromosomes.   Between them, they form a new cell that once again contains the correct number of chromosomes and all the hereditary characteristics of the species.

55. Enzymes are proteins that act as catalysts by generally lowering the activation energy of specific biochemical reactions.   With the assistance of enzymes, these chemical reactions proceed at high speed at normal body temperature.

56. Enzymes are usually specific for one particular reaction because the substrate (substance acted upon by the enzyme) usually fits exactly into a small part of the enzyme to form an intermediate enzyme–substrate complex.   Most substrates do not fit any other enzyme.

57. Polypeptides are numbered starting with the N–terminal amino acid.

N–terminal   tyr   C–terminal   val

58.   1   2   3   4   5
tyr – gly – his – phe – val

59. The correct statements are a, b, c, d, e, f, j, k, m, o, r, s, u, v, y

(g)   Fats have a higher percentage of saturated fatty acids than oils.

(h)   Fats are digested in the intestine and converted to glycerol, mono and diglycerides, and fatty acids.

(i)   Arachidonic, linoleic, and linolenic acids are the three essential fatty acids required by humans.

(l)   The double helix of DNA is tied together by hydrogen bonds between the nitrogen bases.

(n)   The amino acid residues in the polypeptide chain are numbered beginning with the N–terminal amino acid.

(p)   Nucleic acids are made up of units called nucleotides.

(q)   The backbone of each strand of DNA is alternating units of deoxyribose and phosphate.

(t)   The subtance acted upon by an enzyme is called a substrate.

(w)   The main function of RNA is protein synthesis.

60. In the lock and key hypothesis the active site of an enzyme exactly fits the complementary-shaped part of a substrate to form an enzyme-substrate reaction complex on the way to forming the products.   In the flexible site hypothesis the enzyme changes its shape to fit the shape of the substrate to form the enzyme-substrate reaction complex.

61. Enzyme specificity is due to the particular shape of a small segment of the enzyme.   This shape fits a complementary shape of the substrate with which the enzyme is reacting.

62. Enzymes act as catalysts for the biochemical reactions.   Their function is to lower the activation energy of these biochemical reactions.